PEANUT AGRICULTURE AND PRODUCTION TECHNOLOGY

Integrated Nutrient Management

PEANUT AGRICULTURE AND PRODUCTION TECHNOLOGY

Integrated Nutrient Management

Zafar Abbas

PG Department of Botany, GF College (M. J. P. Rohilkhand University), Shahjahanpur–242001, UP, India

Arvind Kumar

Department of Botany, Swami Brahmanand Degree College, Agampur, Shahabad, Hardoi–241124, UP, India

Anoop Kumar

PG Department of Botany, GF College (M. J. P. Rohilkhand University), Shahjahanpur–242001, UP, India

APPLE ACADEMIC PRESS

Apple Academic Press Inc.
3333 Mistwell Crescent
Oakville, ON L6L 0A2 Canada

Apple Academic Press Inc.
9 Spinnaker Way
Waretown, NJ 08758 USA

© 2018 by Apple Academic Press, Inc.

First issued in paperback 2021

Exclusive worldwide distribution by CRC Press, a member of Taylor & Francis Group
No claim to original U.S. Government works

ISBN 13: 978-1-77-463634-3 (pbk)
ISBN 13: 978-1-77-188613-0 (hbk)

Library and Archives Canada Cataloguing in Publication

Abbas, Zafar (Senior associate professor), author
Peanut agriculture and production technology : integrated nutrient management / Zafar Abbas (PG Department of Botany, GF College (M. J. P. Rohilkhand University), Shahjahanpur–242001, UP, India), Arvind Kumar (Department of Botany, Swami Brahmanand Degree College, Agampur, Shahabad, Hardoi–241124, UP, India), Anoop Kumar (PG Department of Botany, GF College (M. J. P. Rohilkhand University), Shahjahanpur–242001, UP, India).
Includes bibliographical references and index.
Issued in print and electronic formats.
ISBN 978-1-77188-613-0 (hardcover).--ISBN 978-1-315-16687-2 (PDF)
1. Peanuts--Nutrition--India. 2. Peanuts--Physiology--India. 3. Peanuts--Harvesting--India. 4. Nitrogen fertilizers--India. 5. Peanut industry--India. I. Kumar, Arvind (Assistant professor of botany) II. Anoop, Kumar (Botany researcher) III. Title.

SB351.P3A23 2017	635'.6596	C2017-905901-7	C2017-905902-5

Library of Congress Cataloging-in-Publication Data

Names: Abbas, Zafar, author.
Title: Peanut agriculture and production technology / authors: Zafar Abbas, Arvind Kumar, Anoop Kumar.
Description: Waretown, NJ : Apple Academic Press, 2017. | Includes bibliographical references and index.
Identifiers: LCCN 2017040136 (print) | LCCN 2017044725 (ebook) | ISBN   9781315166872 (ebook) | ISBN 9781771886130 (hardcover : alk. paper)
Subjects: LCSH: Peanuts.
Classification: LCC SB351.P3 (ebook) | LCC SB351.P3 A23 2017 (print) | DDC   633.3/68--dc23
LC record available at https://lccn.loc.gov/2017040136

Apple Academic Press also publishes its books in a variety of electronic formats. Some content that appears in print may not be available in electronic format. For information about Apple Academic Press products, visit our website at **www.appleacademicpress.com** and the CRC Press website at **www.crcpress.com**

CONTENTS

ABOUT THE AUTHORS

Zafar Abbas, PhD
Senior Associate Professor and Chairman,
PG Department of Botany, GF College (M. J. P.
Rohilkhand University), Shahjahanpur, Uttar Pradesh,
India

Zafar Abbas, PhD, is currently Senior Associate Professor and Chairman in the PG Department of Botany at GF College (M. J. P. Rohilkhand University) in Shahjahanpur, Uttar Pradesh, India. He has 40 years of research experience in plant and crop physiology, with a specialization in plant nutrition. He has attended several national and international seminars and conferences. At present, eight PhD students have completed their doctorate degrees under his supervision. Dr. Abbas is a life member and member of editorial boards of several Indian and international journals and societies, and he has authored a book and has published over 30 articles.

Arvind Kumar, PhD
Assistant Professor of Botany, S. B. N. Degree
College, Shahabad, Hardoi, Uttar Pradesh, India

Arvind Kumar, PhD, is Assistant Professor of Botany at the S. B. N. Degree College in Shahabad Hardoi, Uttar Pradesh, India. He is the recipient of an R. G. N. Research Fellowship from the University Grants Commission, New Delhi. He has five years of teaching experience and has attended several national and international seminars and conferences. He has published several articles in international peer-reviewed journals.

Anoop Kumar, PhD
Botany Researcher, Botany Department, GF College,
Shahjahanpur, Uttar Pradesh, India

Anoop Kumar, PhD, is a botany researcher in the PG Department of Botany at GF College, Shahjahanpur, UP, India. Dr. Kumar has 10 years of research experience and has attended several national and international seminars and conferences. His research has focused on sugarcane nutrition. He has published 15 research papers. He received his PhD in Botany from M. J. P. Rohilkhand University, Bareilly, Uttar Pradesh, India.

LIST OF ABBREVIATIONS

mg	microgram
%	percentage
*	significant
AICRPO	All India Coordinated Research Project on Oilseeds
ANOVA	analysis of variance
BDH	British Drug House
BHC	benzenehexachloride
CBS	cultured bacterial solution
C.D.	critical difference
cm	centimeter
CPE	cummulative pan evaporation
D	optical density
DAP	days after planting
DAS	days after sowing
DDW	double distilled water
D.F.	degree of freedom
DMA	dry matter accumulation
DOR	Directorate of Oilseeds Research
EC	electrical conductivity
F Test	F-test is a method to compare variance of two different sets of values
FLD	field legistative directorate
FYM	farm yard manure
gm	gram
ha	hectare
I	irrigation water
ICOC	Indian Central Oilseeds Committee
ICRISAT	International Crop Research Institute for Semi Arid Tropics
IU	International Units

K	potassium
l	litre
LAI	leaf area index
m	oil content
m	meter
m_0	weight of seed sample
mg	milligram
ml	milliliter
MSS	mean sum of the square
NARC	National Agricultural Research Centres
N.S.	non-significant
O.D.	optical density
ODC	Oilseeds Development Council
P	phosphorus
PIRCOM	Project for Intensification of Regional Research on Cotton, Oilseeds and Millets
PSB	phosphate solubilizing bacteria
PSM	phosphorus solubilizing microorganisms
RDF	recommended dose of fertilizer
S Em	standard error mean
S.S.	sum of the square
SPM	sulphinated pressmud
sq m	square meter
SSP	single super phosphate
t	ton
TDM	total dry matter
TMO	Technology Mission on Oilseeds
USDA	U.S. Department of Agriculture
V	final volume of 80% acetone
w	fresh weight of the sample

FOREWORD

The groundnut is the fourth most important source of edible oil and an equally important source of protein in the world. In the Indian context it is the most important oil seed crop, being grown over 55 lakh hectare area with a production of over 61 lakh tons. In view of the shortage of vegetable oils in the country, it is one crop that can meet our requirement of oil seeds, thereby offsetting volatility in vegetable oil prices. This crop has a great potential in a country like India where it can be grown under different agroclimates including marginal lands, which are available in abundance in the country. It is all the more important to expand its cultivation in view of its high energy conversion efficiency and high biological nitrogen fixation potential that will replenish the fast declining carbon reserves of our soils. What is more striking is its ability to not only sustain the adverse impact of climate change but also improve productivity despite high temperatures and rising CO_2 levels.

Although India is the second largest producer of groundnut in the world, its productivity is far below China and USA. Within India also, its productivity varies from state to state, being highest in Tamil Nadu and lowest in Karnataka and Uttar Pradesh. The differences are almost three times in the productivity of these states. Therefore, our immediate task should be to analyze the reasons for these wide differences in productivity and try to narrow them down by overcoming the factors involved. Even though we have come a long way in expanding the area and production of groundnut in the country by way of developing many high-yielding varieties, production, and protection technologies, much needs to be done on many fronts, which include developing short duration high-yielding varieties, identifying new plant types with higher harvest index, disease and pest forecasting, improving water saving technologies, and looking at yield gap analyses and ways and means to bridge them. Many more such need-based interventions at regional and local level are needed as well.

I am happy to note that a very sincere effort has been made by the authors (Dr. Zafar Abbas and coworkers) in generating a vast pool of data on agrotechniques, which are always the bottleneck in improving the productivity of any crop, including groundnut at local level. The information generated and presented in this book will prove useful to the growers of the region and elsewhere of similar agroclimatic zones of India and abroad. The methodology used in generating this information is robust, and the researchers will findit useful in carrying out their research work on any fieldcrop. Therefore, the book is going to be very useful for both the line departments, postgraduate and PhD students, and the scientists. My sincere appreciation to the authors for their excellent work.

—*Dr. Bir Pal Singh*
Central Potato Research Institute
(Indian Council of Agricultural Research)
Shimla–171001, Himachal Pradesh, India

PREFACE

The present book, *Peanut Agriculture and Production Technology: Integrated Nutrient Management*, focuses on agricultural techniques and integrated nutrient management of peanuts (*Arachis hypogaea* L.). Peanuts are the second most important oil crop of India, occupying 5.7 million hectares, with an average production of 0.8 ton/ha, which is 23.5% of India's total oil seed production. Worldwide annual production of shelled peanuts was 42 million metric tons in 2014. It is the world's fourth most important source of edible oil and the third most important source of vegetable protein.

This book caters to the needs of students in advanced-level programs in agriculture, horticulture and allied sciences. It will also be valuable for agricultural scientists (plant and crop physiology), agronomists, soil scientists, farm owners, managers, researchers, etc., and for the agricultural sector as a whole.

Normal physiology is maintained under ideal environmental conditions. However, plants seldom exist under just the right conditions. Usually something is lacking, and often several factors are far from ideal. Because of the fact of competition, plants often live at the limit of their capability to survive one or more sub-normal conditions. This creates considerable hungers in the organism, which reacts by various biochemical and physiological mechanisms to surpass the sub-optimal effects by managing the development of plants in proper suitable conditions for special needs. In this context peanut growing in the rainy (kharif) season under different sources and methods of nitrogen applications as well as influenceof different sowing dates and population densities to harvest its full yield potential (Part A) as well as on balanced nutrition integrated with organic manures in groundnut *Arachis hypogaea* L. (Part B) to enhance yielding ability, have been considered in the present book.

I am extremely thankful to, firstly, my parents, my wife S. Sultana Rizvi, my children Shuja, Farah, Unsa, and Faraz, and my sweet granddaughter Sahrish, for their co-operative and moral supports. In addition, Apple Academic Press must be praised for their active work and support for our effort.

We are sure that readers of this book—scientists, researchers, professors etc.—will find it interesting and useful.

Above all, I am thankful to the Almighty.

—Dr. Zafar Abbas

CHAPTER 1

INTRODUCTION

CONTENTS

1.1 AGRO-TECHNIQUES: INTRODUCTION

Groundnut (*Arachis hypogaea* L.), belonging to family Fabaceae, is an important annual legume in the world for oilseed, food, and animal feed. Being a soil nitrogen-fixing crop, it is safe, cheap, and eco-friendly for the soil environment. It is a good source of vegetable oil and protein for humans. It is an exhaustive crop and therefore, its proper crop husbandry has become a must. In this chapter, we will consider the significance and importance of the crop with different sources and methods of nitrogenous fertilizers as well as sowing dates and population densities (seed rates), based on intensive field experimentations under local conditions.

1.2 FOCUS ON CROP PRODUCTION

In India (during 2013–2014), groundnut (*Arachis hypogaea* L.) is culti-vated in an area of 5.53 m ha with a production of 9.67 million tons and a productivity of 1,750 kg/ha. The seeds contain 45–50% edible oil, which is used in the preparation of vanaspati, soaps, cosmetics, and cold creams in addition to common cooking oil. According to Das (1997), groundnut oil is good from both the nutritive and culinary perspective, as it contains high oleic acid (40–50%) and linoleic acid (25–35%). The remaining 50% of the kernel has high quality protein (on an average 25.3%), which is about 1.3 times higher than meat, 2.5 times higher than eggs; carbohydrates (6–24.9%); and minerals and vitamins. The productivity is low as compared to the United States and China due to imbalanced fertilization, uncertainty of monsoons, and poor cultural practices adopted by farmers (Kumar, 2012). Being a leguminous crop efficacy of different sources and methods of nitrogen fertilizers as well as sowing dates and population densities were considered to harvest its optimum yielding ability, as these affect different growth and physiological parameters (Kumar, 2012).

Groundnut is an oilseed crop with 40–50% oil content and grown in nearly 100 countries. The remaining seed protein can be used as a meal for food or feed (25–30% protein) (Ahmed et al., 2007). Groundnut is the thirteenth most important food crop of the world (Hatam and Abbasi, 1994). It is the world's fourth most important source of edible oil and third most important source of vegetable protein. Groundnut seeds contain high quality edible oil, easily digestible protein, and carbohydrates.

Groundnuts are rich in essential nutrients. Ozcan (2010) explained that among the common cooking and salad oil, the groundnut oil contains 46% of monosaturated fats (primarily oleic acid), 33% of polyunsaturated fats (primarily linoleic acid), and 17% of saturated fats (primarily palmitic acid).

The remaining 50% of the kernel has high quality protein (on an average 25.3%), which is about 1.3 times higher than meat and 2.5 times higher than eggs; carbohydrates (6–24.9%); and minerals and vitamins. The *United States Department of Agriculture* (2014) provides the following data:

Nutritional value of groundnut per 100 g (3.5 oz)

Energy	2,385 kJ (570 kcal)
Carbohydrates	21 g
Sugars	0.0 g
Dietary fiber	9 g

Fat	48 g
Saturated	7 g
Monosaturated	24 g
Polyunsaturated	16 g
Protein	25 g
Tryptophan	0.2445 g
Threonine	0.859 g
Isoleucine	0.882 g
Leucine	1.627 g
Lysine	0.901 g
Methionine	0.308 g
Cystine	0.322 g
Phenylalanine	1.300 g
Tyrosine	1.020 g
Valine	1.052 g
Arginine	3.001 g
Histidine	0.634 g
Alanine	0.997 g
Aspartic acid	3.060 g
Glutamic acid	5.243 g
Glycine	1.512 g
Proline	1.107 g
Serine	1.236 g
Vitamins	
Thiamine (B1)	0.6 mg
Riboflavin (B2)	0.3 mg
Niacin (B3)	12.9 mg
Pantothenic acid	1.8 mg
Vitamin B6	0.3 mg
Folate (B9)	246 ug

Vitamin C	0.0 mg
Vitamin E	6.6 mg
Minerals	
Calcium	62 mg
Iron	2 mg
Magnesium	184 mg
Manganese	2.0 mg
Phosphorus	336 mg
Potassium	332 mg
Zinc	3.3 mg
Other constituents	
Water	4.26 g

The United States Department of Agriculture (2014), http://ndb.nal.usda.gov/ndb/foods/show/4831.

Units: ug = microgram, mg = milligram, IU = International Units.

It is grown on 26.4 million ha worldwide with a total production of 36.1 million metric tons, and an average productivity of 1.4 metric tons per hectare (FAO, 2004). China leads in production, followed by India as shown in the following table (Anonymous, 2015).

Rank	Countries	Production
1	China	17.0
2	India	9.5
3	Nigeria	3.0
4	United States	1.9
5	Myanmar	1.4
Total	World	46

In India, groundnut is grown on 5.7 million ha with a production of 4.7 million tons and an average productivity of 0.8 tons/ha during the rainy season. Anantpur district of Andhra Pradesh is the largest producer of groundnut with 0.74 million ha of area under cultivation (ICRISAT, 2008). Groundnut haulms (vegetative plant parts) provide excellent hay for feeding livestock, being rich in protein and have better palatability and digestibility than other fodder (Singh et al., 2010). Apart from health care, source of fodder, and improved soil fertility, production of groundnut provides a means of livelihood to scores of people. It is also eco-friendly. Moreover, these crops can be intercropped compatibly with sugarcane (a long-term exhaustive crop), as they increase microbial biomass, control weeds and increase nitrogen (N) availability (Suman et al., 2006). This intercropping will also increase the margin of profit and the economic security of small and marginal farmers in sugarcane-dominant districts and state (Singh et al., 2006).

Modern agriculture is based, to a large extent, on the efficient use of inputs and maximization of yields through proper manipulation of some of the physical factors (agro-techniques) within a given environment. The productivity of groundnut in India is low as compared to the United States and China due to imbalanced fertilization, uncertainty of monsoons, and poor cultural practices adopted by farmers (Kumar, 2012). Keeping in view the importance of this kharif (summer) crop, the author decided to test this crop from the point of view of (i) impact of sources and methods of nitrogen application, and (ii) improvement of agro-techniques (seed rate and dates of sowing) to lend ample scope for intercropping, to promote the needs of farmers, and to maintain long-term soil health—a study not formerly done under local conditions. Two field experiments were conducted on groundnut, between 2008 and 2011, with the following aims and objectives:

1. To study the effect of different sources and methods of nitrogen application on growth, leaf nutrient (NPK) content, pod yield, and oil of groundnut (*Arachis hypogaea* L.).
2. To study the effect of different seed rate/population density, dates of sowing and their interaction on growth, leaf nutrient (NPK) content, pod yield, and oil of groundnut (*Arachis hypogaea* L.).

Both experiments were properly replicated, and the data analyzed statistically. The conclusions drawn have been discussed in light of the findings of other researchers in this book.

1.3   PRESENT SCENARIO OF THE RESEARCH

1.3.1   REVIEW OF LITERATURE

1.3.1.1   Nitrogen Use in *Arachis hypogaea L.*

Nitrogen recoveries from applied fertilizers can be increased by modifying crop geometry and sources and timings of nitrogen application in most of the crops. The importance of nitrogen as an essential plant nutrient need not be overemphasized. It is an integral part of a large variety of biological molecules, most of which are essential for one physiological process or the other. Thus, amino acids, structural proteins, enzymes, coenzymes, purines, pyrimidines, nucleosides, nucleotides, phosphoproteins, lipoproteins, nucleic acids, and protoplasm itself all have nitrogen as an integral element in them (Salisbury and Ross, 2000; Devlin and Witham, 2005; Verma, 2007). In fact, it is difficult to name a basic property of cells in plants that does not at some point make contact with their nitrogen metabolism (Steward and Durzan, 1965).

Groundnut (*Arachis hypogaea* L.) is the major kharif oilseed crop of India. It is an exhaustive crop and removes a large amount of nutrients from the soil. Higher seed yield is mainly attributed to greater biomass, pod weight, 100-seed weight, shelling percent, and harvest index in different genotypes (Singh et al., 2010), for which nitrogen is of prime importance (Kumar et al., 1994; Varalakshmi et al., 2005).

Singh and Ahuja (1985) made a field experiment with groundnut var. T 64 at RBS College research farm, Bichpuri, Agra in the kharif of 1978 and 1979. They reported that nitrogen increased dry matter production at all the growth stages. Application of 25 kg N/ha produced significantly higher dry matter and increased the nutrient uptake as compared to inoculum alone. The nitrogen also increased the yield of nuts and kernels significantly in both years. Inoculation alone gave the lowest yield. Thus, application of 25 kg N/ha proved superior to inoculation alone.

Selamat and Gardner (1985) reported that groundnut (*Arachis hypogaea* L.) normally uses symbiotically fixed N_2 for growth and development, but limitations inherent in the symbionts or environment may result in N deficiencies. Limitations can be corrected by N fertilization, which is a common practice in commercial groundnut production. Ball et al. (1983) found that N application increased total plant and pod yields of "argentine" and "florigiant" peanut cultivars in both 1978 and 1979, but reduced the harvest index, i.e., the vegetative component was increased more in proportion to the fruit. Wynne et al. (1979) showed an increase in both biomass yield (47%) and pod yield (32%) of peanut cultivars on a sandy soil fertilized with N at monthly intervals. On a sandy loam soil, N fertilization reduced seed yield; 100-seed weight of inoculated plants improved with increased seed yield compared with increased seed yield in inoculated plants (Reddy et al., 1981). Groundnut growers commonly believe that high yields require N fertilization at seeding despite effectiveness of biological N2 fixation in peanut. Nitrogen fertilization may improve seedling vigor, competition against weeds, and correct possible deficiencies in growth.

Selamat and Gardner (1985) further reported the effects of N fertilization on growth, N uptake, and N partitioning in nodulating and non-nodulating groundnut genotypes during 1982 in the field and a greenhouse at the University of Florida, Gainesville. Three lines, florunner (FL), early bunch (EB), and a non-nodulating line (M4^{-2}) were grown in the field on a Lakeland sandy soil (hyperthermic coated thermic Quartzipsamment) at four N rates (0, 6, 12, and 24 gm^{-2} N) and in pots in the greenhouse at N rates at a factor of 0.05, 0.50, 0.95, and 1.40 N-strength of Hoagland no. 2 solution, modified to adjust the other nutrients at 0.50 strength. In the field experiment, only the biomass crop growth rate (CGR) and leaf area index (LAI) of all genotypes were significantly increased by N rates. An LAI of 3.0 was achieved at 60 days after planting (DAP) when 6 g m^{-2} N or more was applied on the nodulating genotypes. For the non-nodulating genotype (M4^{-2}), 120 days at 6 g m^{-2} N or 85 days at 24 g m^{-2} N were required to achieve LAI of 3.0. The vegetative and reproductive responses of M4^{-2} to N increments were highly positive and essentially linear. Pod and seed yields of nodulated genotypes in field-grown plants were not significantly influenced by N fertilization. Pod yield was increased in pot-grown plants in the greenhouse, but nodule number and weight were reduced. A genotype × N rates interaction for plant component yield, leaf N concentration, and N content was observed due primarily to the differential response to N between the nodulating and non-nodulating lines. Photoassimilate partition to pods was generally reduced by N fertilization, probably due to vegetative growth stimulation. FL cultivar biologically fixed more N2 than EB; the amount in FL exceeding the seed requirement (12.2 m^{-2}) if fertilizer N rate was low. These studies suggest that the practice of N fertilization of nodulated groundnut does not increase pod or seed yield and is economically counter-productive.

Hadwani and Gundalia (2005) indicated that the pod and haulm yield of groundnut significantly increased with increasing levels of NP K (nitrogen, phosphorus, and potassium). The application of NP increased the pod and haulm yield by 5.7, 14.0, and 26.4 and 34.9% with the application of R_{100} and R_{150}, respectively over control (R_0). Oil and protein yields of groundnut also increased significantly over control. Increase in the yields with increasing level of N and P could be ascribed to the overall improvement in plant growth and vigor as this plays an important role in plant metabolism, resulting in better yield attributes and yields. The results are in accordance with Patel et al. (1994).

The effect of N, P, and K significantly increased total uptake of NPK by groundnut. The highest total uptake of N (155.9), P (14.54), and K (29.4) kg ha^{-1} was noted under the treatment combination $R_{150} \times K_{100}$. There was a synergistic effect of N and P × K interaction on the total uptake of NPK, which was obvious and directly related to the increasing yield as well as their concentration in the plants. Patra et al. (1995) reported that total uptake of NPK increased due to the combined application of various levels of NPK. The application of NP significantly increased the protein content, 100-seed weight, and shelling percentage over control (R_0).

1.3.1.2 Seed Rate/Population Density in *Arachis hypogaea* L.

Change in inter- and intra-row spacings affect the crop growth and yield variations due to acute competition between the crop plants. Fewer plants approach more closely their potential yield. Humans while trying to grow successful, healthy field crops, create such an intense competition that the individual plants are in qualitative terms markedly subnormal (Hasan and Abbas, 2007b).

Singh and Ahuja (1985) studied (*Arachis hypogaea* L.) cv. T 64 at RBS College research farm, Bichpuri, Agra in kharif 1978, 1979 for dry matter accumulation, oil content, and nutrient uptake as affected by fertilizers and plant density. Among three plant densities (111 thousand plants/ha, 148,000 plants/ha), significantly higher yields of nuts and kernels were recorded at a medium density of 148,000 plants/ha. The dry matter production/plant increased with each reduction in plant density at all the stages.

A lowest density of 111 thousand plants/ha recorded significantly higher dry matter/plant than medium and high plant densities.

At high plant density, plants competed from early stage on; thus, growth rate was reduced which became progressively more marked as competition intensified.

Azu and Tanner (1978) also reported corroborative results. At the 40 days stage, plant density also influenced the nodule weight significantly. Low density produced significantly higher dry weight of nodules as compared to medium and high densities. Increasing inter-plant competition by increasing plant density adversely affected the nodule size and nodule weight. This inverse relationship of plant density to nodule weight was also observed by Bhan and Misra (1970). The highest yield were recorded with low density (111 thousands plant/ha) but the highest with medium plants density (148 thousand plant/ha). In both cases, further increase in density caused significant reduction in yield.

Regarding nutrient uptake, Singh and Ahuja (1985) noted that among the plant densities, medium density of 148 thousand plants/ha recorded significantly higher total uptake than high and low densities. A significant increase in total uptake of nutrients was recorded with each increase. The protein and oil percentages of seed significantly decreased by every increase of plant densities in both the years in *Arachis hypogaea* L. (groundnut).

1.3.1.3   Date of Sowing in *Arachis hypogaea* L.

Amongst oilseed crops, groundnut occupies a predominant position in terms of acreage and total production. Under assured irrigation conditions, advance planting (prior to onset of monsoon) enhances productivity of this crop owing to least exposure to adverse situations (Gill and Kumar, 1995; Kumar et al., 2003). However, there is need to assess feasibility of different planting dates and its impact on productivity under prevailing agro-climatic conditions. Change in date of sowing induces the crop growth period to face different temperatures at different stages automatically and naturally. Temperature plays an important role in all aspects of plant growth and development (Went, 1953; Hasan and Abbas, 2007b).

Ketring (1984) reported that optimum temperature for growth of most groundnut (*Arachis hypogaea* L.) genotypes is near 30°C. During the 1980 growing season in the United States, temperatures ranged from 35 to 40°C for many days in the groundnut producing regions, and drought occurred. Experiments were initiated to determine the effect of temperature separate from drought on groundnut development. Plants were grown in controlled environments at 30/25°C, 12/12 h light/dark temperature to obtain a population of plants with uniform development. Measurements of individual leaf areas, leaf dry weights, and seedling height were made 21 days after planting to establish plant growth state. Then temperature treatments, based on temperatures expected to occur during the summer groundnut growing season in the south west United States, were begun. Temperature during the dark period was held constant at 22°C for all treatments. Temperatures during the light period were 30, 32, and 35°C. The 35°C treatment decreased individual leaf areas and dry weight at both 63 and 91 days after planting. Plants harvested at 91 days after planting showed reduced total leaf area; stem elongation was decreased in two experiments at 35°C. The highest temperature treatment significantly reduced the number of subterranean pegs. Mature seed weight was reduced and significantly so in two of three experiments, once for each cultivar tested. Thus, a temperature of 35°C was shown to have an inhibitory effect on groundnut development even when plants were grown in well-watered conditions. An evaluation of eight peanut genotypes showed genotypic variation in response to the 35°C temperature treatment. Thus, selection of peanuts with improved heat tolerance may be possible.

Talwar et al. (1999) found among abiotic factors, high temperature is one of the major constraints to adaptation of groundnut (*Arachis hypogaea* L.) in tropical and subtropical areas. The aims of their study were (i) to evaluate three genotypes (ICG 1236, ICGS 44, and Chico) of groundnut for their heat acclimation potential (HAP), and (ii) examine whether the growth, yield, and photosynthetic responses of these genotypes to temperature related to HAP. HAP was defined as the change in leaf tolerance based on plasmalemma thermostability at 40 to 60°C measured by electrolyte leakage after acclimation at 35/30°C day/night temperature. Initially, plants were raised in a glasshouse maintained at 25/25°C day/night temperature. One half of the plants were shifted to another glasshouse maintained at 35/30°C after the appearance of the third leaf. Heat killing time (HKT), defined as the time required to cause 50% relative injury, indicated that the three genotypes acclimated to high temperature stress, with significant variations in HAP. All genotypes maintained greater vegetative growth and higher photosynthetic rates when grown under the higher temperature regime and genetic differences in photosynthetic rate were related to HKT. However, the higher temperature regime affected the reproductive growth adversely by increasing flower abortion and decreasing seed size. Differences in chlorophyll fluorescence and membrane thermo-stability between growth temperatures were found only after incubating the leaf tissue at temperatures of 50°C or higher. Genetic differences in HAP were small and unrelated to growth differences.

Kumar et al. (2003) conducted a field experiment during the kharif seasons of 1997 and 1998 at the Agricultural Research Station, Durgapura, Jaipur, to assess impact of advance sowing on groundnut (*Arachis hypogaea* L.) productivity. Amongst yield components, sowing dates failed to bring about perceptible variation in pods setting index, kernel pod, shelling percent, and seed index, but number of pods and their weight/plant were significantly improved under advance planting on June 1 over the rest. The crop sown on June 1 recorded significantly higher productivity in terms of pod, kernel, and haulm yield (30.81, 20.38, and 48.78 q/ha), representing an increase of 21.58, 25.88, and 21.50% and 40.23, 49.85, and 48.20% over June 16 and July 1 sowing, respectively.

Carley et al. (2008) found that determining when to dig groundnut (*Arachis hypogaea* L.) is complicated because of its indeterminate growth habit. Pod mesocarp color is often used as an indicator of pod maturation. However, this process is time consuming and is usually based on a relatively small subsample of pods from groundnut fields. Research was conducted in 2003–2005 to determine if reflectance of the groundnut canopy, using multispectral imaging (350–2500 nm), could be used as an indicator of pod maturation. The cultivars VA 98R and NC-V II were planted beginning in early May through early June in each year with reflectance and the percentage of pods at optimum maturity (percentage of pods with brown or black mesocarp color) determined in mid-September. The highest yield observed for VA 98R across the three years of the experiment was noted when peanut was planted in mid-May rather than early or late May or when planted in early June when groundnut was dug based on optimum pod maturity using pod mesocarp color. Pod yield for the cultivar NC-V II did not differ when comparing planting dates. For cultivar VA 98R, Pearson's correlations were significant in all bandwidth categories except the normalized difference vegetation index (NDVI) when reflectance was compared with percentage of mature pods. Reflectance for NC-V II was not significant for any of the correlations though significant differences in the percentage of mature pods were noted in mid-September when comparing planting dates. These data suggest that canopy reflectance could potentially aid in predicting pod maturation, but more research is needed to determine feasibility of this approach.

1.3.2   CONCLUDING REMARKS AFTER LITERARY REVIEW

It may be concluded that, in view of groundnut (*Arachis hypogaea* L.) being an important cash crop and new genotypes being available with improved agro-techniques, the literature is not only meager but also patchy. Moreover, as this kharif crop may act as an income-generating crop for household nutrition and economic security, various aspects of its cultivation merits intensive study.

1.4   METHODOLOGY AND PLAN OF WORK

1.4.1   MATERIAL AND METHODS

1.4.1.1   General

The experiments included in the present study were conducted in the field during the kharif (summer) season of the years 2008–2010 at the Botanical

Garden and Agricultural Farm, G.F. College, Shahjahanpur, Uttar Pradesh. TG 37-A variety of groundnut (*Arachis hypogaea* L.) was selected for the purpose. The study is comprised of two field experiments, and the detail of the experiments with performed and prevailing conditions are given in the following sections.

1.4.1.2 Agro-Climatic Conditions, Soil Characteristics, and Locality

Shahjahanpur is located at latitude 27°53′ N, longitude 79°4′ E, and at altitude 154.53 meters above sea level (Figure 1.1). It has the semi-arid and sub-tropical climate of the Tarai region with hot dry summers and cold winters. The average precipitation in the years 2008, 2009, and 2010 was 412.18 mm, 560 mm, and 748.7 mm, respectively. Most of this annual precipitation was received during the two months of June and July. The temperature touched 38.5, 39.1, and 40.2°C during the crop growth period; although, it occasionally fell to as low as 8.2, 6.7, and 9.3°C, respectively, in the three years. The meteorological data for the period of the investigation, recorded by the Meteorological Observatory, U.P. Council of Sugarcane Research, Shahjahanpur, U.P. have been presented (Figures 1.2–1.7).

After preparation of the field, soil samples were collected randomly at a depth of 15 cm from each of the experimental field in the three years and analyzed for physico-chemical properties. The data are given in Table 1.1.

1.4.2 PREPARATION OF THE FIELD

Before beginning each field experiment, the field was thoroughly ploughed to ensure maximum aeration. Plots of 5.0 sq.m. size were prepared in which uniform recommended basal dose of farmyard manure 20 q/ha was applied in each experiment area to ensure better soil environment. Standard agricultural practices recommended for the groundnut crop by state agricultural universities/research institutes were followed.

FIGURE 1.1   Showing location and details of Shahjahanpur district (west U.P.).

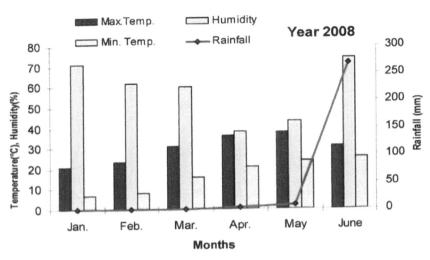

FIGURE 1.2 Showing temperature, relative humidity and rainfall (monthly basis) from January to June 2008.

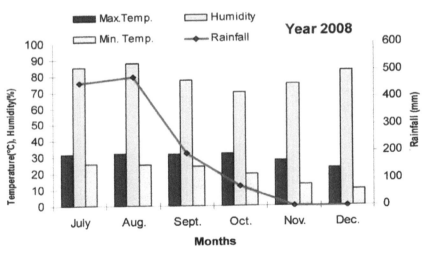

FIGURE 1.3 Showing temperature, relative humidity and rainfall (monthly basis) from July to December 2008.

1.4.3 PROCUREMENT OF SEEDS

Authentic seeds of groundnut (*Arachis hypogaea* L.) variety TG 37-A were obtained from C.S.A. University of Agriculture and Technology, Kanpur (U.P.).

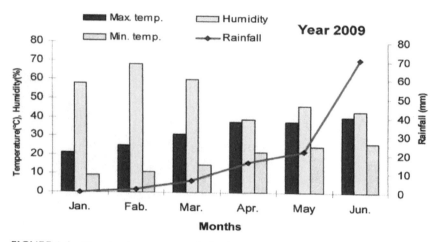

FIGURE 1.4 Showing temperature, relative humidity and rainfall (monthly basis) from January to June 2009.

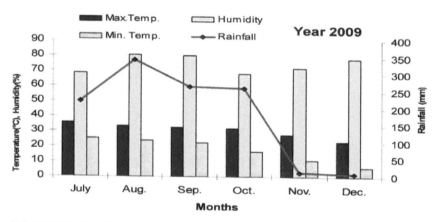

FIGURE 1.5 Showing temperature, relative humidity and rainfall (monthly basis) from July to December 2009.

1.4.4 EXPERIMENT 1

This first field experiment was conducted on groundnut (*Arachis hypogaea* L.) var. TG 37-A during the kharif (summer) season of 2009 in a factorial randomized block design. The seeds were sown with 30 × 10 cm row to plant distances and treated with 3 g thiram per kg before sowing to make them disease free. The physico-chemical soil analysis of the

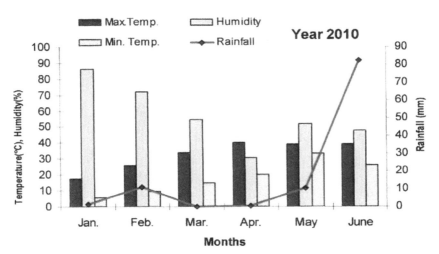

FIGURE 1.6 Showing temperature, relative humidity and rainfall (monthly basis) from January to June 2010.

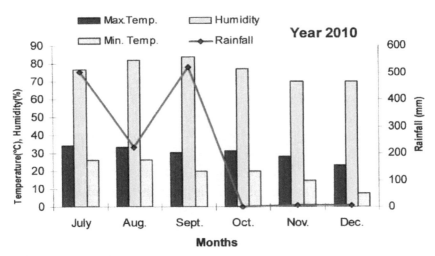

FIGURE 1.7 Showing temperature, relative humidity and rainfall (monthly basis) from July to December 2010.

experiment field is given in the Table 1.1. The seeds were sown on June 3, 2009. This experiment on groundnut was conducted to study the effect of nitrogen sources (S1 = urea, S2 = ammonium sulfate, S3 = calcium ammonium nitrate, and S4 = diammonium phosphate) @ 50 kg N/ha applied with different methods of applications (T1 = full soil, T2 = full foliar, T3

TABLE 1.1   Physico-Chemical Analysis of Soil of the Experimental Fields 2008 and 2010

Characteristics	Experiment 1 (2009)	Experiment 2 (2010)
Texture	Sandy loam	Sandy loam
Particle size distribution:		
Sand	70.0	65.0
Silt	15.0	20.0
Clay	15.0	15.0
pH	7.10	7.50
E.C. dS/m	0.30	0.40
Organic Carbon (%)	0.25	0.20
Available N (kg/ha)	196.0	270.0
Available P (kg/ha)	14.10	12.70
Available K (kg/ha)	110.00	115.00
Calcium Carbonate	Nil	Nil

= 2/3 soil + 1/3 foliar, T4 = 1/3 soil + 2/3 foliar, T5 = 1/4 soil + 3/4 foliar, and T6 = 3/4 soil + 1/4 foliar) on growth characters; leaf chlorophylls and carotenoid content; leaf nutrient (NPK) content sampled at pre-flowering (50 days), flowering (70 days), and post-flowering (90 days) stages, as well as at harvest; yield characteristics; and pod and oil yield (%) of each treatment replicated thrice. An aqueous 1% nitrogen solution was sprayed foliarly 40 days after sowing, where required. A uniform dose of 40 kg, each of P and K was applied during field preparation through calcium super phosphate and muriate of potash along with farmyard manure. The crop was irrigated when required. The harvesting was done on October 20, 2009. The summary of treatments is given in Table 1.2.

1.4.5   EXPERIMENT 2

This second field experiment was conducted on *Arachis hypogaea* L. (groundnut) variety TG 37-A to determine the effect of agro-techniques, different seed rates (row to plant distances) and dates of sowing (at one month intervals) and their interaction in a factorial randomized design in the kharif (summer) season of 2010. Each treatment was replicated thrice.

TABLE 1.2   Model of Summary of the Treatments of Experiment 1

Methods of Nitrogen Application	Sources of N Fertilizer			
	Urea	Ammonium sulphate	Calcium ammonium nitrate	Diammonium phosphate
T_1 = Full soil				
T_2 = Full foliar				
T_3 = 2/3 Soil + 1/3 Foliar				
T_4 = 1/3 Soil + 2/3 Foliar				
T_5 = 1/4 Soil + 3/4 Foliar				
T_6 = 3/4 Soil + 1/4 Foliar				

The physico-chemical analysis of the experimental field (soil) is already provided in Table 1.1. The seeds were sown on May 5, June 5, July 5, August 5, and September 5, 2010. Pre-sowing, they were treated with 3 gm thiram per kg to make them disease free. The crop was assessed by growth characteristics; leaf chlorophylls and carotenoid contents; leaf nutrient (NPK) contents sampled at pre-flowering (50 days), flowering (70 days), and post-flowering (90 days) stages, as well as at harvest; yield characteristics; and pod and oil yield (%). The summary of the treatments is given in Table 1.3. The crop was harvested on maturity at different timings according to the schedule of changed crop growth periods on September 5, October 5, November 5, November 25, and December 15, 2010,

TABLE 1.3   Model of Summary of the Treatments of Experiment 2

Seed Rate/Plant Density (row to plant)	Dates of sowing				
	5 May	5 June	5 July	5 August	5 September
T_1 = 40 x 10 cm					
T_2 = 40 x 5 cm					
T_3 = 30 x 10 cm					
T_4 = 30 x 5 cm					
T_5 = 20 x 10 cm					
T_6 = 20 x 5 cm					

respectively. The crop was irrigated when required and standard agricultural practices were followed according to the crop needs.

1.4.6 SAMPLING TECHNIQUES

Three plants from each bed of the experimental plot of 2.0 × 2.5 sq. m. were uprooted randomly at various growth stages (50, 70, and 90 days approx.) for the study of various growth characteristics and leaf nutrient (NPK) content of plants. The plants for studying pod yield and yield attributes with seed oil content (%) were also randomly collected at harvest. The yield per hectare was calculated. The pod yield was noted after a week of harvest and drying in the sun. Seeds (kernel) were stored for assessing the oil content.

1.4.7 GROWTH CHARACTERS

The germination percentage was noted a fortnight after sowing. To assess the growth performance, the following parameters were chosen in the crop for the study sampled at pre-flowering, flowering, and post-flowering (about 50, 70, and 90 days to sowing) stages.
- Fresh weight (g/plant);
- Leaf number/plant;
- Plant height (cm);
- Root length (cm);
- Dry weight (g/plant).

After recording the growth characteristics, the plant sample of each treatment were kept in an oven at 80°C to estimate dry weight (till constant weight was achieved). Leaf powder was made from dried leaves and used to determine leaf NPK content. Chlorophylls and carotenoids were estimated in fresh leaves.

1.4.8 QUANTITATIVE ANALYSIS OF CHLOROPHYLLS AND CAROTENOIDS (ARNON, 1949)

Fresh leaves were taken from plants sampled at pre-flowering, flowering, and post-flowering stages for chlorophylls and carotenoid content.

Chloroplasts are dynamic entities of green cells in which photosynthesis takes place. Light energy is absorbed by chloroplasts and starch is synthesized with CO_2 and H_2O, O_2 being the by-product. The whole sustenance of a plant depends upon these green pigments. Chlorophyll a and b have typical absorption spectra of solar radiation. Maximum peak of chlorophyll a observed in the blue–violet region is at 429 nm and in the red region at 660 nm; chlorophyll b has its maximum interception at 453 nm and 642 nm. These chlorophylls are best known and are found in all autotrophic organisms. Chlorophyll a is usually blue-green and Chlorophyll b is yellow-green. The empirical formula of chlorophyll a molecule is $C_{55}H_{72}O_5N_4Mg$. However, the absorption spectra of chlorophyll in vivo may be quite different.

The method stated by Arnon (1949) is described here since it is accessible to all researchers. One can estimate these pigments if a spectrophotometer is available in the laboratory.

1.4.8.1 Material Required

The materials required are: (i) Acetone, BDH/Analar; (ii) Whatman paper No. 1; (iii) Funnel stands; (iv) Mortar and pestle; (v) Volumetric flasks (25 ml.); and (vi) Spectronic-20/spectrophotometer.

1.4.8.2 Procedure

About 80% acetone was prepared with 250 mg of fresh leaf material, excluding midribs, was accurately weighed on an electronic monopan balance, and transferred to a mortar and pestle for grinding. 10 ml of 80% acetone was added in it, and the leaves were thoroughly macerated. A homogenous paste was made. The chlorophyll extract was poured in a funnel having Whatman paper No. 1 and collected in a volumetric flask. Every care was taken to collect all the extract from the mortar. A green extract was gradually obtained by adding acetone (80%) 5 ml at a time. Three to four washings were given, and extraction continued until the leachate became colorless. The final volume should not exceed 25 ml. In the experiment, the volume was made to 25 ml with 80% acetone. Since the extract was subjected to photo-oxidation, it was kept away from direct sunlight and stored in a refrigerator. The optical density of the chlorophyll extract was recorded on a Spectronic-20 colorimeter using 480, 510, 645, 652, and 663 nm wavelengths. Leaf samples were analyzed in duplicate for safety.

The following formulas were used for calculating chlorophyll a, b and carotenoids:

Observation and Data:
1. Chlorophyll a (mg/g) = 12.7 ($D \times 663$) – 2.69 ($D \times 645$)
2. Chlorophyll b (mg/g) = 22.9 ($D \times 645$) – 4.68 ($D \times 633$)
3. Carotenoids = 7.6 ($D \times 480$) – 1.49 ($D \times 510$)

where, D = optical density.

1.4.9 LEAF NUTRIENT (NPK) CONTENTS

Leaf analysis is now well established as a tool for assessing the nutritional requirements of a particular crop plant (Lundegardh, 1951). It signifies a definite relationship between the content of the nutrients in the leaves and vegetative growth; the index values of various nutrients at different stages of a crop plant show a clear picture for the assessment of nutritional requirements. On the other hand, the amount or concentration of various nutrients present in the soil does not always show a direct correlation with plant growth, as many factors, such as the area of absorbing surface, antagonism and synergism, soil fixation, etc., influence their availability to the physiology of the plant itself apart from the several edaphic factors that may also play an important role in the process of absorption.

After measuring dry weight, dried leaves were taken from the sample, finely powdered and passed through a 72 mesh screen. The powder was stored in polythene bags, labeled and kept for analyses. Dried leaves powder were also taken from the plants sampled at the different stages mentioned earlier, for nutrient analyses.

The following leaf nutrients were estimated on the basis of percent dry weight.

1. Leaf nitrogen content (%);
2. Leaf phosphorus content (%); and
3. Leaf potassium content (%).

1.4.9.1 Digestion of Plant Samples

About 100 mg of the oven dried leaf powder of each sample was transferred to a 50 ml Kjeldahl flask to which 2 ml sulfuric acid was added. The contents of the flask was heated on a temperature-controlled assembly for about 2 hours to allow complete reduction of nitrates present in the plant material by the organic matter itself. As a result, the contents of the flask turned black. After cooling the flask for about 15 minutes, 0.5 ml of 30% hydrogen peroxide was added drop by drop and the solution was heated again till the color of the solution changed from black to light yellow. After cooling for about 30 minutes, an additional 3–4 drops of 30% hydrogen peroxide was added, followed by heating for another 15 minutes. The addition of 30% hydrogen peroxide, followed by heating was repeated if the content of the flask do not become colorless. The digested peroxide material was transferred from the Kjeldahl flask to a 100 ml volumetric flask with three washings, each with 5 ml double distilled water (DDW). Volume of the volumetric flask was made up to the mark with DDW (100 ml).

1.4.9.1.1 Estimation of Nitrogen

Estimation of nitrogen was carried out using the method followed by Lindner (1944). A 10 ml aliquot of the digested material was taken in a 50 ml volumetric flask. To this, 2 ml of 2.5 N sodium hydroxide and 1 ml of 10% sodium silicate solution was added to neutralize excess of acid and to prevent turbidity, respectively. The volume of the solution was made up to the mark with distilled water. In a 10 ml graduated test tube, a 5 ml aliquot of this solution was taken and 0.5 ml of Nessler's reagent was added. The final volume was made up with distilled water. The content of the tube was allowed to stand for 5 minutes for maximum color development. The solution was transferred to a colorimetric tube and optical density (OD) was read at 525 nm with the help of a spectrophotometer.

1.4.9.1.1.1 Standard Curve for Nitrogen

About 50 mg ammonium sulfate was dissolved in 1 l DDW. From this solution, 0.1, 0.2, 0.3, 0.4, 0.5, 0.6, 0.7, 0.8, 0.9, and 1.0 ml was pipetted to ten different test tubes. The solution in each test tube was diluted to 5 ml with DDW. In each test tube, 0.5 ml Nessler's reagent was added. After 5 minutes, the optical density was red at 525 nm on a spectrophotometer. A blank was run with each set of determination. The standard curve was prepared using different dilutions of ammonium sulfate solution versus optical density and with the help of the standard curve, the amount of nitrogen present in the sample was determined.

1.4.9.1.2   Estimation of Phosphorus

The total amount of phosphorus in the sulfuric acid peroxide digested material was estimated by the method followed by Fiske and Subbarow (1925). A 5 ml aliquot was taken in a 10 ml graduated test tube and 1 ml of molybdic acid (2.5% ammonium molybdate in 10 N H_2SO_4), was added carefully, followed by the addition of 0.4 ml of 1-amino-2-naph-thol 1–4 sulfonic acid. The color turned blue. DDW was used to make up the volume to 10 ml. The solution was shaken for 5 minutes and then transferred to a colorimetric tube. The optical density was read at 620 nm on a spectrophotometer. A blank was used simultaneously with each determination.

1.4.9.1.2.1   Standard curve for phosphorus

About 351 mg potassium dihydrogen orthophosphate was dissolved in sufficient DDW to which 10 ml of 10 N sulfuric acid was added and the final volume was made up to 1 l with DDW. From this solution, 0.1, 0.2, 0.3, 0.4, 0.5, 0.6, 0.7, 0.8, 0.9, and 1.0 ml was taken in ten different test tubes. The solution in each test tube was diluted to 5 ml with DDW. In each test tube, 1 ml molybdic and 0.4 ml 1-amino-2-naphthol 1–4 sulfonic acid was added. After 5 minutes, the optical density was read at 620 nm on spectrophotometer. A blank was also run with each set of determination. The standard curve was prepared using different dilutions of potassium dihydrogen orthophosphate solution versus optical density and with the help of the standard curve, the amount of phosphorus present in the sample was determined.

1.4.9.1.3 Estimation of Potassium

Potassium was estimated with the help of a flame photometer. After adjusting the filter for potassium in the flame photometer, 10 ml peroxide digested material was run. A blank was also run side by side with each set of determination.

1.4.9.1.3.1 Standard curve for potassium

Potassium chloride (1.91 gm) was dissolved in 100 ml of DDW, of which 1 ml solution was diluted to 1 l. The resulting solution was a 10 ppm solution of potassium. From this 10 ppm solution, 1, 2, 3, 4, 5, 6, 7, 8, 9, and 10 ml solution was transferred to 10 vials separately. The solution in each vial was diluted to 10 ml with DDW. The diluted solution of each vial was run separately. A blank was also run with each set of determination. The standard curve was prepared using different dilutions of potassium chloride solution versus reading on the scale of the flame photometer. The amount of potassium present in the sample was determined with the help of this standard curve.

1.4.10 YIELD CHARACTERISTICS IN GROUNDNUT (ARACHIS HYPOGAEA L.)

The plants were allowed to grow to maturity. The following yield and oil content characteristics were studied for assessment at the time of harvest (about 115 days after sowing).
  i.   Weight/plant (g);
  ii. Number of pod/plant;
  iii. Weight of pod/plant (g);
  iv. Pod yield (kg/ha);
  v.   Total oil content (%).

1.4.11 SEED OIL EXTRACTION IN GROUNDNUT (ARACHIS HYPOGAEA L.)

Oil content was extracted by the Soxhlet method using petroleum ether as the extractant. The oil yield was expressed in percentage. About 25 g of the crushed seed were transferred to a Soxhlet apparatus and sufficient quantity of petroleum ether was added. The apparatus was kept on a hot water bath running at 60°C for about 6 hours for complete extraction of the oil. The petroleum ether from the extracted oil was evaporated. The extracted oil was expressed as a percentage by mass of seeds and was calculated by the following formula:

$$m/m_0 \times 100$$

where, m = mass in g of oil; m_0 = seed sample in (g).

1.4.12 STATISTICAL ANALYSIS

All the experiments were statistically designed and the data collected were subjected to statistical analysis by adopting analysis of variance (ANOVA) techniques according to Panse and Sukhatme (1967). F-tests were carried out in which the error due to replicates was also determined. In cases where the F-test was noted to be significant at the 5% level of probability, critical difference (CD) was calculated. The models of the analysis of variance (ANOVA) for each of the two experiments performed are given in Tables 1.4 and 1.5.

1.5 DATA TABULATION AND ANALYSIS

1.5.1 EXPERIMENTAL RESULTS

1.5.1.1 Experiment 1

The first experiment was conducted in 2009 on groundnut (*Arachis hypogaea* L.) var. TG 37-A during the kharif (summer) season in a factorial randomized design to study the effect of nitrogen sources (S1 = urea, S2 = ammonium sulfate, S3 = calcium ammonium nitrate and S4 = diammonium phosphate) @ 50 kg N/ha applied using different methods (T1 = full soil, T2 = full foliar, T3 = 2/3 soil + 1/3 foliar, T4 = 1/3 soil + 2/3 foliar, T5 = 1/4 soil + 3/4 foliar, and T6 = 3/4 soil + 1/4 foliar) on growth

TABLE 1.4 ANOVA for the Effect of Sources and Methods of Nitrogen Application

Source of variation value	D.F.	S.S.	M.S.S.	"F"
Replications	2			
Treatments	5			
Varieties	3			
Treatments X varieties	15			
Error	46			
Total	71			

TABLE 1.5 ANOVA for the Effect of Dates of Sowing and Population Density

Source of variation value	D.F.	S.S.	M.S.S.	"F"
Replications	2			
Treatments	5			
Varieties	4			
Treatments X varieties	20			
Error	58			
Total	89			

characters; leaf chlorophylls and carotenoid content; leaf nutrient (NPK) content sampled at pre-flowering (50 days), flowering (70 days), and post-flowering (90 days) stages as well as at the harvest yield characteristics, pod yield, and oil yield (%). The important results are summarized in Tables 1.7–1.45.

1.5.1.1.1 Growth Characteristics

The effect of various methods of nitrogen application, sources of nitrogenous fertilizers and their interaction effect were studied at 50, 70, and 90 days of sowing on growth characteristics (germination%, fresh weight, leaf number, plant height, root length, dry weight, leaf chlorophyll a, leaf chlorophyll b, and leaf carotenoid). The results in brief are described in the following subsections

1.5.1.2 Germination%

The effect on germination% as a result of different methods of nitrogenous application, nitrogen sources, and interaction effect was noted to be non-significant (Tables 1.6 and 1.7).

1.5.1.2.1 Fresh Weight (g/plant)

The fresh weight (g/plant) was significantly affected by different methods of nitrogen application in groundnut (*Arachis hypogaea* L.) at different growth stages (Tables 1.8–1.10). At 50 days, the highest value was noted in full soil application statistically equal to 1/3 soil + 2/3 foliar application followed by 3/4 soil + 1/4 foliar method of application (Table 1.8). At 70 days and 90 days, 2/3 soil + 1/3 foliar seemed to work better (Tables 1.9 and 1.10) followed by full soil application. Full foliar application treatment also proved poorest for fresh weight in groundnut.

TABLE 1.6   Effect of Different Sources and Methods of Nitrogen Application on Germination % (One Week After Sowing) of *Arachis hypogaea* L. (Groundnut)

Methods of nitrogen application	Source nitrogen fertilizer				
	Urea	Ammonium sulphate	Calcium ammonium nitrate	Diammonium phosphate	Mean
T1 = Full soil	60.43	63.10	60.80	61.00	61.33
T2 = Full foliar	62.10	63.20	61.30	61.80	62.10
T3 = 2/3 Soil + 1/3 Foliar	61.30	63.30	62.90	62.30	62.45
T4 = 1/3 Soil + 2/3 Foliar	61.90	63.10	62.10	62.30	62.35
T5 = 1/4 Soil + 3/4 Foliar	62.00	63.10	63.10	61.30	62.37
T6 = 3/4 Soil + 1/4 Foliar	61.80	61.00	62.30	60.30	61.35
Mean	61.58	62.80	62.08	61.50	
	Nitrogen fertilizer (A)		Nitrogen application (B)		A × B
S.Em.±	0.542		0.621		1.130
CD at 5%	1.130		2.101		4.302
Significance	NS		NS		NS

TABLE 1.7 Effect of Different Sources and Methods of Nitrogen Application on Germination (%) of *Arachis hypogaea* L. (Groundnut)

Methods of nitrogen application	Source nitrogen fertilizer				
	Urea	Ammonium sulphate	Calcium ammonium nitrate	Diammonium phosphate	Mean
T_1 = Full soil	83.34	86.67	86.67	86.00	85.67
T_2 = Full foliar	86.67	86.00	86.00	86.00	86.17
T_3 = 2/3 Soil + 1/3 Foliar	86.00	86.00	83.33	86.34	85.42
T_4 = 1/3 Soil + 2/3 Foliar	86.67	86.00	83.32	83.33	84.83
T_5 = 1/4 Soil + 3/4 Foliar	86.66	86.00	83.34	86.67	85.67
T_6 = 3/4 Soil + 1/4 Foliar	86.00	86.65	86.00	86.66	86.33
Mean	85.89	86.22	84.78	85.83	
	Nitrogen fertilizer (A)		**Nitrogen application (B)**		**A × B**
S.Em.±	0.646		0.791		1.583
CD at 5%	1.840		2.254		4.508
Significance	NS		NS		NS

TABLE 1.8 Effect of Different Sources and Methods of Nitrogen Application on Fresh Weight (g/plant) of *Arachis hypogaea* L. (Groundnut) at 50 Days

Methods of nitrogen application	Source nitrogen fertilizer				
	Urea	Ammonium sulphate	Calcium ammonium nitrate	Diammonium phosphate	Mean
T_1 = Full soil	177.00	162.34	172.34	187.67	174.84
T_2 = Full foliar	84.34	73.00	61.00	67.34	71.42
T_3 = 2/3 Soil + 1/3 Foliar	158.67	169.00	194.00	154.33	169.00
T_4 = 1/3 Soil + 2/3 Foliar	151.66	131.67	123.00	144.66	137.75
T_5 = 1/4 Soil + 3/4 Foliar	157.33	134.00	120.34	152.34	141.00
T_6 = 3/4 Soil + 1/4 Foliar	161.67	144.00	148.00	169.67	155.84
Mean	148.45	135.67	136.45	146.00	
	Nitrogen fertilizer (A)		**Nitrogen application (B)**		**A × B**
S.Em.±	3.14		3.84		7.69
CD at 5%	8.94		10.95		21.90
Significance	*		**		**

TABLE 1.9   Effect of Different Sources and Methods of Nitrogen Application on Fresh Weight (g/plant) of *Arachis hypogaea* L. (Groundnut) at 70 Days

Methods of nitrogen application	Source nitrogen fertilizer				
	Urea	Ammonium sulphate	Calcium ammoni- um nitrate	Diam- monium phosphate	Mean
T_1 = Full soil	284.34	262.67	276.00	283.34	276.59
T_2 = Full foliar	206.00	184.00	190.67	198.00	194.67
T_3 = 2/3 Soil + 1/3 Foliar	284.67	270.66	286.00	271.66	278.25
T_4 = 1/3 Soil + 2/3 Foliar	288.66	278.34	284.33	284.34	283.92
T_5 = 1/4 Soil + 3/4 Foliar	193.34	188.00	187.00	188.67	189.25
T_6 = 3/4 Soil + 1/4 Foliar	263.67	255.00	274.00	256.00	262.17
Mean	253.45	239.78	249.67	247.00	

	Nitrogen fertilizer (A)	Nitrogen application (B)	A × B
S.Em.±	3.55	4.35	8.70
CD at 5%	10.11	12.38	24.77
Significance	NS	**	NS

TABLE 1.10   Effect of Different Sources and Methods of Nitrogen Application on Fresh Weight (g/plant) of *Arachis hypogaea* L. (Groundnut) at 90 Days

Methods of nitrogen application	Source nitrogen fertilizer				
	Urea	Ammonium sulphate	Calcium ammoni- um nitrate	Diam- monium phosphate	Mean
T1 = Full soil	304.67	394.66	382.34	308.00	347.42
T2 = Full foliar	248.66	231.34	212.31	292.67	246.25
T3 = 2/3 Soil + 1/3 Foliar	313.00	393.33	396.34	360.33	365.75
T4 = 1/3 Soil + 2/3 Foliar	321.67	324.32	305.33	379.34	332.67
T5 = 1/4 Soil + 3/4 Foliar	326.66	327.00	309.00	363.32	331.50
T6 = 3/4 Soil + 1/4 Foliar	311.34	315.33	319.67	318.33	316.17
Mean	304.33	331.00	320.83	337.00	

	Nitrogen fertilizer (A)	Nitrogen application (B)	A × B
S.Em.±	3.51	4.30	8.60
CD at 5%	10.00	12.24	24.49
Significance	**	**	**

The response of nitrogenous fertilizers at 50 and 90 days was noted to be significant (Tables 1.11–1.13). Diammonium phosphate proved to be best for fresh weight.

The interaction effect was also noted significant only at 50 and 70 days (Tables 1.11–1.13). At 50 days, full soil application of diammonium phosphate was best in value and statistically equal to 3/4 soil + 1/4 foliar application. At 90 days, 2/3 soil + 1/3 foliar application of ammonium sulfate proved to be best followed by 1/3 soil + 2/3 foliar (Table 1.11).

1.5.1.2.2 Leaf Number

Leaf number was significantly affected due to various methods of nitrogen application (Tables 1.11–1.13). At 50 days, full soil application was best and statistically equal to 1/3 soil + 2/3 foliar as well as 2/3 soil + 1/3 foliar. At 70 days, full soil application was highest in value and statistically equal to 2/3 soil + 1/3 foliar. At 90 days, 3/4 soil + 1/4 foliar was best and statistically equal to 2/3 soil + 1/3 foliar. The effect of full foliar application on leaf production was poorest at all the growth stages (Tables 1.11–1.13).

TABLE 1.11 Effect of Different Sources and Methods of Nitrogen Application on Leaf Number/Plant of *Arachis hypogaea* L. (Groundnut) at 50 Days

Methods of nitrogen application	Source nitrogen fertilizer				
	Urea	Ammonium sulphate	Calcium ammonium nitrate	Diammonium phosphate	Mean
T_1 = Full soil	67.34	72.34	61.00	66.00	66.67
T_2 = Full foliar	48.00	39.33	39.00	43.66	42.50
T_3 = 2/3 Soil + 1/3 Foliar	61.67	79.66	55.34	78.67	68.84
T_4 = 1/3 Soil + 2/3 Foliar	62.67	69.00	68.67	78.33	69.67
T_5 = 1/4 Soil + 3/4 Foliar	64.00	66.34	66.66	51.34	62.09
T_6 = 3/4 Soil + 1/4 Foliar	68.67	75.00	74.33	66.90	71.23
Mean	62.06	66.95	60.83	64.15	

	Nitrogen fertilizer (A)	Nitrogen application (B)	A × B
S.Em.±	0.98	1.20	2.41
CD at 5%	2.80	3.44	6.88
Significance	**	**	**

TABLE 1.12 Effect of Different Sources and Methods of Nitrogen Application on Leaf Number/Plant of *Arachis hypogaea* L. (groundnut) at 70 Days

Methods of nitrogen application	Source nitrogen fertilizer				
	Urea	Ammonium sulphate	Calcium ammoni-um nitrate	Diam-monium phosphate	Mean
T_1 = Full soil	148.00	167.00	159.67	149.66	156.08
T_2 = Full foliar	82.67	72.34	73.00	94.00	80.50
T_3 = 2/3 Soil + 1/3 Foliar	145.66	169.88	149.66	151.67	154.22
T_4 = 1/3 Soil + 2/3 Foliar	132.34	124.67	141.67	151.34	137.51
T_5 = 1/4 Soil + 3/4 Foliar	139.00	125.34	126.33	121.33	128.00
T_6 = 3/4 Soil + 1/4 Foliar	150.33	162.00	154.67	142.00	152.25
Mean	133.00	136.87	134.17	135.00	
	Nitrogen fertilizer (A)		Nitrogen application (B)		A × B
S.Em.±	2.44		2.98		5.97
CD at 5%	6.94		8.50		17.01
Significance	NS		**		**

TABLE 1.13 Effect of Different Sources and Methods of Nitrogen Application on Leaf Number/Plant of *Arachis hypogaea* L. (groundnut) at 90 Days

Methods of nitrogen application	Source nitrogen fertilizer				
	Urea	Ammonium sulphate	Calcium ammoni-um nitrate	Diam-monium phosphate	Mean
T_1 = Full soil	96.34	84.67	73.32	89.66	86.00
T_2 = Full foliar	85.00	86.65	86..34	89.67	87.11
T_3 = 2/3 Soil + 1/3 Foliar	87.34	102.34	85.00	96.00	92.67
T_4 = 1/3 Soil + 2/3 Foliar	93.32	82.66	88.67	83.34	87.00
T_5 = 1/4 Soil + 3/4 Foliar	79.67	75.66	71.00	73.00	74.83
T_6 = 3/4 Soil + 1/4 Foliar	92.34	95.00	99.00	85.00	92.84
Mean	89.00	87.83	83.40	86.11	
	Nitrogen fertilizer (A)		Nitrogen application (B)		A × B
S.Em.±	1.28		1.57		3.15
CD at 5%	3.67		4.49		8.99
Significance	*		**		**

As far as the interaction effect was concerned, at 50 days, 2/3 soil + 1/3 foliar application of ammonium sulfate was best and statistically equal to 2/3 soil + 1/3 foliar application of diammonium phosphate.

1.5.1.2.3   Plant Height (cm)

Plant length was significantly affected due to different methods of nitrogen application (Tables 1.14–1.16). At 50 days, plants that had 3/4 soil + 1/4 foliar application were tallest, and statistically equal to 2/3 soil + 1/3 foliar as well as 1/3 soil + 2/3 foliar (Table 1.14). At 70 days, 2/3 soil + 1/3 foliar was best followed by 3/4 soil + 1/4 foliar. At 90 days, 2/3 soil + 1/3 foliar was best and statistically equal to 1/3 soil + 2/3 foliar as well as 3/4 soil + 1/4 foliar treatments. The effect of full foliar applied nitrogen was poorest at all growth stages (Tables 1.14–1.16).

The interaction effect was significant at 70 days only for plant height. The highest value was noted in 2/3 soil + 1/3 foliar application of

TABLE 1.14   Effect of Different Sources and Methods of Nitrogen Application on Plant Height (cm) of *Arachis hypogaea* L. (Groundnut) at 50 Days

Methods of nitrogen application	Source nitrogen fertilizer				
	Urea	Ammonium sulphate	Calcium ammonium nitrate	Diammonium phosphate	Mean
T₁ = Full soil	54.00	62.00	56.33	61.00	58.33
T₂ = Full foliar	53.34	52.67	51.66	52.00	52.42
T₃ = 2/3 Soil + 1/3 Foliar	58.66	63.66	58.67	61.00	60.50
T₄ = 1/3 Soil + 2/3 Foliar	53.67	53.34	54.00	56.90	54.48
T₅ = 1/4 Soil + 3/4 Foliar	52.34	52.33	51.33	52.34	52.09
T₆ = 3/4 Soil + 1/4 Foliar	61.00	62.34	59.66	59.67	60.67
Mean	55.50	57.72	55.28	57.15	
	Nitrogen fertilizer (A)		Nitrogen application (B)		A × B
S.Em.±	0.84		1.03		2.07
CD at 5%	2.41		2.95		5.91
Significance	NS		**		NS

TABLE 1.15 Effect of Different Sources and Methods of Nitrogen Application on Plant Height (cm) of *Arachis hypogaea* L. (Groundnut) at 70 Days

Methods of nitrogen application	Source nitrogen fertilizer				
	Urea	Ammonium sulphate	Calcium ammonium nitrate	Diammonium phosphate	Mean
T_1 = Full soil	59.00	67.34	67.00	67.00	65.09
T_2 = Full foliar	61.34	60.67	67.00	67.66	64.17
T_3 = 2/3 Soil + 1/3 Foliar	76.00	79.34	78.00	71.33	76.17
T_4 = 1/3 Soil + 2/3 Foliar	69.67	65.00	69.66	65.33	67.42
T_5 = 1/4 Soil + 3/4 Foliar	68.00	64.00	64.33	64.33	65.17
T_6 = 3/4 Soil + 1/4 Foliar	69.33	70.67	70.34	70.66	70.25
Mean	67.22	67.84	69.39	67.72	
	Nitrogen fertilizer (A)		**Nitrogen application (B)**		**A × B**
S.Em.±	0.70		0.86		1.73
CD at 5%	2.01		2.47		4.94
Significance	NS		**		**

TABLE 1.16 Effect of Different Sources and Methods of Nitrogen Application on Plant Height (cm) of *Arachis hypogaea* L. (Groundnut) at 90 Days

Methods of nitrogen application	Source nitrogen fertilizer				
	Urea	Ammonium sulphate	Calcium ammonium nitrate	Diammonium phosphate	Mean
T1 = Full soil	84.67	87.34	83.34	87.00	85.59
T2 = Full foliar	84.65	85.32	88.66	89.67	87.08
T3 = 2/3 Soil + 1/3 Foliar	88.66	91.34	84.33	90.34	88.67
T4 = 1/3 Soil + 2/3 Foliar	88.67	86.00	86.67	89.33	87.67
T5 = 1/4 Soil + 3/4 Foliar	81.34	82.34	84.00	80.00	81.92
T6 = 3/4 Soil + 1/4 Foliar	85.33	86.67	85.00	90.34	86.84
Mean	85.55	86.50	85.33	87.78	
	Nitrogen fertilizer (A)		**Nitrogen application (B)**		**A × B**
S.Em.±	0.91		1.12		2.24
CD at 5%	2.61		3.19		6.39
Significance	NS		**		NS

ammonium sulfate, statistically equal to 2/3 soil + 1/3 foliar application of urea (Table 1.15).

1.5.1.2.4 Root Length (cm)

Root length of groundnut was significantly affected due to different methods of nitrogen application at all growth stages (Tables 1.17–1.19). At 50 days, the highest value was found in 2/3 soil + 1/3 foliar, statistically equal to 3/4 soil + 1/4 foliar as well as 1/3 soil + 2/3 foliar. Similar trends in root length were noted at 70 days and 90 days as a result of different treatments (Tables 1.18–1.19). The effect of full foliar nitrogen application was poorest at all growth stages (Tables 1.17–1.19).

The response of ammonium sulfate was better when compared to other sources; this was followed by diammonium phosphate. Among the various interactions, 2/3 soil + 1/3 foliar application of ammonium sulfate was best. The same combination was also maximum at 70 and 90 days (Tables 1.18–1.19).

TABLE 1.17 Effect of Different Sources and Methods of Nitrogen Application on Root Length (cm) of *Arachis hypogaea* L. (Groundnut) at 50 Days

Methods of nitrogen application	Source nitrogen fertilizer				
	Urea	Ammonium sulphate	Calcium ammonium nitrate	Diammonium phosphate	Mean
T1 = Full soil	7.00	7.33	8.00	8.67	7.75
T2 = Full foliar	6.34	6.34	6.50	6.17	6.34
T3 = 2/3 Soil + 1/3 Foliar	8.50	9.67	8.66	8.00	8.71
T4 = 1/3 Soil + 2/3 Foliar	8.34	8.66	8.17	9.34	8.63
T5 = 1/4 Soil + 3/4 Foliar	8.84	8.84	8.17	8.50	8.59
T6 = 3/4 Soil + 1/4 Foliar	8.84	8.83	8.84	8.17	8.67
Mean	7.98	8.28	8.06	8.14	
	Nitrogen fertilizer (A)		Nitrogen application (B)		A × B
S.Em.±	0.071		0.087		0.174
CD at 5%	0.202		0.248		0.496
Significance	*		**		**

TABLE 1.18 Effect of Different Sources and Methods of Nitrogen Application on Root Length (cm) of *Arachis hypogaea* L. (Groundnut) at 70 Days

Methods of nitrogen application	Source nitrogen fertilizer				
	Urea	Ammonium sulphate	Calcium ammonium nitrate	Diammonium phosphate	Mean
T_1 = Full soil	9.50	9.80	9.80	9.20	9.58
T_2 = Full foliar	8.00	8.80	8.80	8.90	8.63
T_3 = 2/3 Soil + 1/3 Foliar	9.70	11.50	10.30	10.80	10.58
T_4 = 1/3 Soil + 2/3 Foliar	10.00	9.50	11.20	11.00	10.43
T_5 = 1/4 Soil + 3/4 Foliar	10.80	9.80	9.70	9.20	9.88
T_6 = 3/4 Soil + 1/4 Foliar	10.50	10.25	11.40	10.00	10.54
Mean	9.75	9.94	10.20	9.85	

	Nitrogen fertilizer (A)	Nitrogen application (B)	A × B
S.Em.±	0.097	0.118	0.237
CD at 5%	0.276	0.338	0.676
Significance	*	**	**

TABLE 1.19 Effect of Different Sources and Methods of Nitrogen Application on Root Length (cm) of *Arachis hypogaea* L. (Groundnut) at 90 Days

Methods of nitrogen application	Source nitrogen fertilizer				
	Urea	Ammonium sulphate	Calcium ammonium nitrate	Diammonium phosphate	Mean
T_1 = Full soil	11.34	11.34	10.84	11.50	11.26
T_2 = Full foliar	12.00	11.67	11.33	12.50	11.88
T_3 = 2/3 Soil + 1/3 Foliar	11.84	13.66	10.83	12.51	12.21
T_4 = 1/3 Soil + 2/3 Foliar	11.50	12.00	11.50	12.17	11.79
T_5 = 1/4 Soil + 3/4 Foliar	11.67	11.84	11.66	12.50	11.92
T_6 = 3/4 Soil + 1/4 Foliar	11.50	12.90	12.84	11.50	12.19
Mean	11.64	12.24	11.50	12.11	

	Nitrogen fertilizer (A)	Nitrogen application (B)	A × B
S.Em.±	0.127	0.155	0.311
CD at 5%	0.361	0.443	0.886
Significance	**	**	**

1.5.1.2.5 Dry Weight (g/plant)

Dry weight of the groundnut plants was significantly affected due to different methods of nitrogen application at all growth stages (Tables 1.20–1.22). At 50 days, maximum dry weight was noted in 2/3 soil + 1/3 foliar, statistically equal to 3/4 soil + 1/4 foliar. At 70 days, the highest value was noted in 2/3 soil + 1/3 foliar for dry weight (Table 1.21). As far as the interaction effect was concerned, at 50 days, 2/3 soil + 1/3 foliar application of ammonium sulfate was best, statistically equal to 2/3 soil + 1/3 foliar application of diammonium phosphate. Almost a similar effect of treatments was also noted at 70 and 90 days (Tables 1.20–1.22).

1.5.1.2.6 Leaf Chlorophyll a (mg/g)

Leaf chlorophyll a content was significantly affected due to different methods of nitrogen application at all growth stages (Tables 1.23–1.25). At 50 days, maximum value was noted in 3/4 soil + 1/4 foliar followed by

TABLE 1.20 Effect of Different Sources and Methods of Nitrogen Application on Dry Weight (g/plant) of *Arachis hypogaea* L. (Groundnut) at 50 Days

Methods of nitrogen application	Source nitrogen fertilizer				
	Urea	Ammonium sulphate	Calcium ammonium nitrate	Diammonium phosphate	Mean
T_1 = Full soil	25.47	32.67	30.54	31.74	30.11
T_2 = Full foliar	27.07	22.66	24.50	29.54	25.94
T_3 = 2/3 Soil + 1/3 Foliar	30.34	34.47	32.20	33.33	32.59
T_4 = 1/3 Soil + 2/3 Foliar	29.74	27.07	29.70	29.60	29.03
T_5 = 1/4 Soil + 3/4 Foliar	30.34	25.14	29.80	31.10	29.10
T_6 = 3/4 Soil + 1/4 Foliar	31.57	32.87	32.17	31.07	31.92
Mean	29.09	29.15	29.82	31.06	
	Nitrogen fertilizer (A)		Nitrogen application (B)		A × B
S.Em.±	0.334		0.410		0.820
CD at 5%	0.953		1.167		2.335
Significance	**		**		**

TABLE 1.21 Effect of Different Sources and Methods of Nitrogen Application on Dry Weight (g/plant) of *Arachis hypogaea* L. (Groundnut) at 70 Days

Methods of nitrogen application	Source nitrogen fertilizer				
	Urea	Ammonium sulphate	Calcium ammonium nitrate	Diammonium phosphate	Mean
T_1 = Full soil	36.87	33.70	35.14	36.67	35.60
T_2 = Full foliar	32.34	27.54	28.00	29.90	29.45
T_3 = 2/3 Soil + 1/3 Foliar	37.87	39.94	37.94	39.00	38.69
T_4 = 1/3 Soil + 2/3 Foliar	37.60	36.34	37.20	38.54	37.42
T_5 = 1/4 Soil + 3/4 Foliar	36.67	38.33	37.20	38.47	37.67
T_6 = 3/4 Soil + 1/4 Foliar	36.90	38.00	39.54	33.00	36.86
Mean	36.38	35.64	35.84	35.93	
	Nitrogen fertilizer (A)		Nitrogen application (B)		A × B
S.Em.±	0.441		0.540		1.081
CD at 5%	1.256		1.539		3.078
Significance	NS		**		**

TABLE 1.22 Effect of Different Sources and Methods of Nitrogen Application on Dry Weight (g/plant) of *Arachis hypogaea* L. (Groundnut) at 90 Days

Methods of nitrogen application	Source nitrogen fertilizer				
	Urea	Ammonium sulphate	Calcium ammonium nitrate	Diammonium phosphate	Mean
T_1 = Full soil	40.84	39.33	39.54	39.14	39.71
T_2 = Full foliar	49.67	46.20	42.20	46.14	46.05
T_3 = 2/3 Soil + 1/3 Foliar	42.47	49.77	39.27	41.54	43.26
T_4 = 1/3 Soil + 2/3 Foliar	44.14	42.67	42.84	45.14	43.70
T_5 = 1/4 Soil + 3/4 Foliar	40.34	40.94	39.94	41.34	40.64
T_6 = 3/4 Soil + 1/4 Foliar	42.30	46.57	47.87	43.60	45.09
Mean	43.29	44.25	41.94	42.82	
	Nitrogen fertilizer (A)		Nitrogen application (B)		A × B
S.Em.±	0.700		0.857		1.715
CD at 5%	1.993		2.441		4.883
Significance	NS		**		*

TABLE 1.23 Effect of Different Sources and Methods of Nitrogen Application on Chlorophyll 'a' (mg/g) of *Arachis hypogaea* L. (Groundnut) at 50 Days

Methods of nitrogen application	Source nitrogen fertilizer				
	Urea	Ammonium sulphate	Calcium ammoni-um nitrate	Diammonium phosphate	Mean
T_1 = Full soil	0.56	0.49	0.52	0.50	0.52
T_2 = Full foliar	0.47	0.49	0.50	0.50	0.49
T_3 = 2/3 Soil + 1/3 Foliar	0.57	0.66	0.45	0.49	0.54
T_4 = 1/3 Soil + 2/3 Foliar	0.43	0.43	0.43	0.46	0.44
T_5 = 1/4 Soil + 3/4 Foliar	0.49	0.45	0.49	0.49	0.48
T_6 = 3/4 Soil + 1/4 Foliar	0.55	0.58	0.58	0.53	0.56
Mean	0.51	0.52	0.50	0.50	
	Nitrogen fertilizer (A)		Nitrogen application (B)		A × B
S.Em.±	0.003		0.004		0.009
CD at 5%	0.010		0.013		0.026
Significance	**		**		**

TABLE 1.24 Effect of Different Sources and Methods of Nitrogen Application on Chlorophyll 'a' (mg/g) of *Arachis hypogaea* L. (Groundnut) at 70 Days

Methods of nitrogen application	Source nitrogen fertilizer				
	Urea	Ammonium sulphate	Calcium ammoni-um nitrate	Diam-monium phosphate	Mean
T_1 = Full soil	0.61	0.60	0.58	0.53	0.58
T_2 = Full foliar	0.52	0.56	0.60	0.55	0.56
T_3 = 2/3 Soil + 1/3 Foliar	0.63	0.74	0.65	0.65	0.67
T_4 = 1/3 Soil + 2/3 Foliar	0.63	0.55	0.52	0.59	0.57
T_5 = 1/4 Soil + 3/4 Foliar	0.60	0.65	0.66	0.55	0.62
T_6 = 3/4 Soil + 1/4 Foliar	0.59	0.65	0.65	0.56	0.61
Mean	0.60	0.63	0.61	0.57	
	Nitrogen fertilizer (A)		Nitrogen application (B)		A × B
S.Em.±	0.006		0.007		0.014
CD at 5%	0.017		0.020		0.041
Significance	**		**		**

TABLE 1.25 Effect of Different Sources and Methods of Nitrogen Application on Chlorophyll 'a' (mg/g) of *Arachis hypogaea* L. (Groundnut) at 90 Days

Methods of nitrogen application	Source nitrogen fertilizer				
	Urea	Ammonium sulphate	Calcium ammoni- um nitrate	Diammoni- um phos- phate	Mean
T_1 = Full soil	0.54	0.56	0.50	0.50	0.53
T_2 = Full foliar	0.52	0.49	0.51	0.48	0.50
T_3 = 2/3 Soil + 1/3 Foliar	0.60	0.65	0.61	0.55	0.60
T_4 = 1/3 Soil + 2/3 Foliar	0.53	0.56	0.54	0.59	0.56
T_5 = 1/4 Soil + 3/4 Foliar	0.45	0.52	0.51	0.57	0.51
T_6 = 3/4 Soil + 1/4 Foliar	0.61	0.59	0.58	0.56	0.59
Mean	0.54	0.56	0.54	0.54	

	Nitrogen fertilizer (A)	Nitrogen application (B)	A × B
S.Em.±	0.0025	0.0031	0.0062
CD at 5%	0.0072	0.0089	0.0178
Significance	**	**	**

2/3 soil + 1/3 foliar. At 70 days and 90 days, maximum value was given by 2/3 soil + 1/3 foliar (Tables 1.24–1.25).

The response of ammonium sulfate was better for leaf chlorophyll a content at all growth stages as compared to other nitrogenous sources (Tables 1.23–1.25).

Among the various interactions, 2/3 soil + 1/3 foliar application of ammonium sulfate was best at all stages followed by 3/4 soil + 1/4 foliar application of ammonium sulfate interaction (Tables 1.23–1.25).

1.5.1.2.7   Leaf Chlorophyll b (mg/g)

Leaf chlorophyll b content was significantly affected due to different methods of nitrogen application at all stages of growth (Tables 1.26–1.28). At 50 days, maximum value was recorded in 2/3 soil + 1/3 foliar followed by 3/4 soil + 1/4 foliar. At 70 days, 3/4 soil + 1/4 foliar was

TABLE 1.26 Effect of Different Sources and Methods of Nitrogen Application on Chlorophyll 'b' (mg/g) of *Arachis hypogaea* L. (Groundnut) at 50 Days

Methods of nitrogen application	Source nitrogen fertilizer				
	Urea	Ammonium sulphate	Calcium ammonium nitrate	Diammonium phosphate	Mean
T_1 = Full soil	0.38	0.31	0.37	0.38	0.36
T_2 = Full foliar	0.31	0.38	0.31	0.35	0.34
T_3 = 2/3 Soil + 1/3 Foliar	0.34	0.48	0.36	0.48	0.42
T_4 = 1/3 Soil + 2/3 Foliar	0.35	0.32	0.39	0.45	0.38
T_5 = 1/4 Soil + 3/4 Foliar	0.30	0.31	0.35	0.30	0.32
T_6 = 3/4 Soil + 1/4 Foliar	0.36	0.45	0.47	0.35	0.41
Mean	0.34	0.38	0.38	0.39	

	Nitrogen fertilizer (A)	Nitrogen application (B)	A × B
S.Em.±	0.003	0.004	0.009
CD at 5%	0.010	0.013	0.026
Significance	**	**	**

TABLE 1.27 Effect of Different Sources and Methods of Nitrogen Application on Chlorophyll 'b' (mg/g) of *Arachis hypogaea* L. (Groundnut) at 70 Days

Methods of nitrogen application	Source nitrogen fertilizer				
	Urea	Ammonium sulphate	Calcium ammonium nitrate	Diammonium phosphate	Mean
T_1 = Full soil	0.31	0.30	0.35	0.35	0.33
T_2 = Full foliar	0.30	0.30	0.35	0.35	0.33
T_3 = 2/3 Soil + 1/3 Foliar	0.31	0.39	0.32	0.38	0.35
T_4 = 1/3 Soil + 2/3 Foliar	0.34	0.36	0.33	0.38	0.35
T_5 = 1/4 Soil + 3/4 Foliar	0.36	0.37	0.30	0.25	0.32
T_6 = 3/4 Soil + 1/4 Foliar	0.38	0.38	0.39	0.38	0.38
Mean	0.33	0.35	0.34	0.35	

	Nitrogen fertilizer (A)	Nitrogen application (B)	A × B
S.Em.±	0.0030	0.0037	0.0075
CD at 5%	0.0087	0.0107	0.0214
Significance	**	**	**

TABLE 1.28 Effect of Different Sources and Methods of Nitrogen Application on Chlorophyll 'b' (mg/g) of *Arachis hypogaea* L. (Groundnut) at 90 Days

Methods of nitrogen application	Source nitrogen fertilizer				
	Urea	Ammonium sulphate	Calcium ammonium nitrate	Diammonium phosphate	Mean
T$_1$ = Full soil	0.25	0.26	0.27	0.26	0.26
T$_2$ = Full foliar	0.20	0.28	0.22	0.25	0.24
T$_3$ = 2/3 Soil + 1/3 Foliar	0.23	0.30	0.25	0.29	0.27
T$_4$ = 1/3 Soil + 2/3 Foliar	0.21	0.22	0.24	0.25	0.23
T$_5$ = 1/4 Soil + 3/4 Foliar	0.21	0.21	0.21	0.23	0.22
T$_6$ = 3/4 Soil + 1/4 Foliar	0.25	0.26	0.27	0.24	0.26
Mean	0.23	0.26	0.24	0.25	

	Nitrogen fertilizer (A)	Nitrogen application (B)	A × B
S.Em.±	0.0034	0.0042	0.0085
CD at 5%	0.0099	0.0121	0.0242
Significance	**	**	**

best, followed by 2/3 soil + 1/3 foliar. At 90 days, 2/3 soil + 1/3 foliar was best, followed by 3/4 soil + 1/4 foliar statistically equal to full soil application (Table 1.28).

The response of ammonium sulfate and diammonium phosphate was better at almost all the growth stages for chlorophyll b content.

The interaction effect of 2/3 soil + 1/3 foliar application of ammonium sulfate was best followed by 3/4 soil + 1/4 foliar application of ammonium sulfate at almost all stages of growth (Tables 1.26–1.28).

1.5.1.2.8 Leaf Carotenoid Content (mg/g)

Leaf carotenoid content in groundnut was significantly affected due to different methods of nitrogen application at all stages of growth (Tables 1.29–1.31). At 50 days, highest value was noted in 3/4 soil + 1/4 foliar followed by 2/3 soil + 1/3 foliar. At 90 days, again 2/3 soil + 1/3 foliar was

TABLE 1.29   Effect of Different Sources and Methods of Nitrogen Application on Carotenoid (mg/g) of *Arachis hypogaea* L. (Groundnut) at 50 Days

Methods of nitrogen application	Source nitrogen fertilizer				
	Urea	Ammonium sulphate	Calcium ammonium nitrate	Diammonium phosphate	Mean
T$_1$ = Full soil	0.12	0.14	0.14	0.14	0.14
T$_2$ = Full foliar	0.18	0.12	0.12	0.11	0.13
T$_3$ = 2/3 Soil + 1/3 Foliar	0.14	0.16	0.14	0.13	0.14
T$_4$ = 1/3 Soil + 2/3 Foliar	0.11	0.12	0.14	0.15	0.13
T$_5$ = 1/4 Soil + 3/4 Foliar	0.13	0.14	0.15	0.14	0.14
T$_6$ = 3/4 Soil + 1/4 Foliar	0.14	0.16	0.16	0.16	0.16
Mean	0.14	0.14	0.14	0.14	

	Nitrogen fertilizer (A)	Nitrogen application (B)	A × B
S.Em.±	0.0025	0.0031	0.0062
CD at 5%	0.0073	0.0089	0.0179
Significance	NS	**	**

TABLE 1.30   Effect of Different Sources and Methods of Nitrogen Application on Carotenoid (mg/g) of *Arachis hypogaea* L. (Groundnut) at 70 Days

Methods of nitrogen application	Source nitrogen fertilizer				
	Urea	Ammonium sulphate	Calcium ammonium nitrate	Diammonium phosphate	Mean
T$_1$ = Full soil	0.50	0.48	0.46	0.45	0.47
T$_2$ = Full foliar	0.45	0.41	0.45	0.46	0.44
T$_3$ = 2/3 Soil + 1/3 Foliar	0.44	0.54	0.48	0.49	0.49
T$_4$ = 1/3 Soil + 2/3 Foliar	0.45	0.44	0.46	0.49	0.46
T$_5$ = 1/4 Soil + 3/4 Foliar	0.39	0.45	0.41	0.38	0.41
T$_6$ = 3/4 Soil + 1/4 Foliar	0.47	0.51	0.51	0.52	0.50
Mean	0.45	0.47	0.46	0.47	

	Nitrogen fertilizer (A)	Nitrogen application (B)	A × B
S.Em.±	0.0089	0.0109	0.0828
CD at 5%	0.0253	0.0310	0.0620
Significance	NS	**	NS

TABLE 1.31   Effect of Different Sources and Methods of Nitrogen Application on Carotenoid (mg/g) of *Arachis hypogaea* L. (Groundnut) at 90 Days

Methods of nitrogen application	Source nitrogen fertilizer				
	Urea	Ammonium sulphate	Calcium ammonium nitrate	Diammonium phosphate	Mean
T1 = Full soil	0.24	0.22	0.22	0.24	0.23
T2 = Full foliar	0.25	0.26	0.24	0.26	0.25
T3 = 2/3 Soil + 1/3 Foliar	0.26	0.33	0.29	0.25	0.28
T4 = 1/3 Soil + 2/3 Foliar	0.23	0.23	0.24	0.29	0.25
T5 = 1/4 Soil + 3/4 Foliar	0.26	0.22	0.27	0.24	0.25
T6 = 3/4 Soil + 1/4 Foliar	0.26	0.29	0.29	0.25	0.27
Mean	0.25	0.26	0.26	0.26	
	Nitrogen fertilizer (A)		Nitrogen application (B)		A × B
S.Em.±	0.0041		0.0051		0.0102
CD at 5%	0.0119		0.0146		0.0292
Significance	NS		**		**

best, statistically equal to 3/4 soil + 1/4 foliar method of nitrogen application. As far as the interaction effect was concerned, 2/3 soil + 1/3 foliar application of ammonium sulfate proved to be better as compared to other combinations (Tables 1.29–1.31).

1.5.1.3   Leaf Nutrient (NPK) Contents

Leaf nutrient (NPK) content was significantly affected by different nitrogen application methods except at 90 days for nitrogen (Table 1.34) and for leaf potassium at 50 days (Table 1.38). The results of leaf nitrogen phosphorus and potassium at the three growth stages are briefly described in the following subsections.

1.5.1.3.1   Leaf Nitrogen (%)

Leaf nitrogen percentage was significantly affected due to different methods of nitrogen application at 50 and 70 days (Tables 1.32–1.34). At 50

TABLE 1.32 Effect of Different Sources and Methods of Nitrogen Application on Leaf Nitrogen (%) of *Arachis hypogaea* L. (Groundnut) at 50 Days

Methods of nitrogen application	Source nitrogen fertilizer				
	Urea	Ammonium sulphate	Calcium ammonium nitrate	Diammonium phosphate	Mean
T_1 = Full soil	3.10	3.09	3.11	3.10	3.10
T_2 = Full foliar	2.89	2.91	2.90	2.92	2.91
T_3 = 2/3 Soil + 1/3 Foliar	3.12	3.41	3.08	3.20	3.20
T_4 = 1/3 Soil + 2/3 Foliar	3.11	3.05	3.11	3.21	3.12
T_5 = 1/4 Soil + 3/4 Foliar	2.98	3.05	3.06	3.02	3.03
T_6 = 3/4 Soil + 1/4 Foliar	3.25	3.30	3.29	3.19	3.26
Mean	3.08	3.14	3.09	3.11	
	Nitrogen fertilizer (A)		Nitrogen application (B)		A × B
S.Em.±	0.061		0.075		0.150
CD at 5%	0.174		0.213		0.427
Significance	NS		*		NS

TABLE 1.33 Effect of Different Sources and Methods of Nitrogen Application on Leaf Nitrogen (%) of *Arachis hypogaea* L. (Groundnut) at 70 Days

Methods of nitrogen application	Source nitrogen fertilizer				
	Urea	Ammonium sulphate	Calcium ammonium nitrate	Diammonium phosphate	Mean
T_1 = Full soil	3.01	2.99	2.98	3.00	3.00
T_2 = Full foliar	2.90	2.90	2.89	2.85	2.89
T_3 = 2/3 Soil + 1/3 Foliar	3.05	3.25	3.02	3.12	3.11
T_4 = 1/3 Soil + 2/3 Foliar	3.01	2.99	2.98	3.11	3.02
T_5 = 1/4 Soil + 3/4 Foliar	2.90	2.96	2.98	2.96	2.95
T_6 = 3/4 Soil + 1/4 Foliar	3.12	3.20	3.22	3.12	3.17
Mean	3.00	3.05	3.01	3.03	
	Nitrogen fertilizer (A)		Nitrogen application (B)		A × B
S.Em.±	0.039		0.047		0.095
CD at 5%	0.111		0.136		0.272
Significance	NS		**		NS

TABLE 1.34 Effect of Different Sources and Methods of Nitrogen Application on Leaf Nitrogen (%) of *Arachis hypogaea* L. (Groundnut) at 90 Days

Methods of nitrogen application	Source nitrogen fertilizer				
	Urea	Ammonium sulphate	Calcium ammonium nitrate	Diammonium phosphate	Mean
T₁ = Full soil	2.86	2.80	2.84	2.86	2.84
T₂ = Full foliar	2.78	2.70	2.78	2.87	2.78
T₃ = 2/3 Soil + 1/3 Foliar	2.81	2.95	2.85	2.90	2.88
T₄ = 1/3 Soil + 2/3 Foliar	2.76	2.70	2.76	2.92	2.79
T₅ = 1/4 Soil + 3/4 Foliar	2.74	2.76	2.78	2.74	2.76
T₆ = 3/4 Soil + 1/4 Foliar	2.86	2.91	2.90	2.80	2.87
Mean	2.80	2.80	2.82	2.85	

	Nitrogen fertilizer (A)	Nitrogen application (B)	A × B
S.Em.±	0.070	0.085	0.171
CD at 5%	0.199	0.244	0.488
Significance	NS	NS	NS

days and 70 days, maximum nitrogen percentage was noted in 3/4 soil + 1/4 foliar method followed by 2/3 + 1/3 foliar treatment.

1.5.1.3.2 Leaf Phosphorus (%)

Leaf phosphorus (%) was significantly affected due to different methods and sources as well as their interactions at all growth stages (Tables 1.35–1.37). Among the different treatments, 3/4 soil + 1/4 foliar was highest at all the stages, followed by 2/3 soil + 1/3 foliar method. The response of diammonium phosphate was best, followed by ammonium sulfate at all growth stages.

As far as the interaction effects were concerned 2/3 soil + 1/3 foliar application of ammonium sulfate proved superior at all the growth stages for leaf phosphorus percentages (Tables 1.35–1.37). The combination 2/3 soil + 1/3 foliar application of diammonium phosphate was also found better as far as leaf phosphorus percentages was concerned.

TABLE 1.35 Effect of Different Sources and Methods of Nitrogen Application on Phosphorus (%) of *Arachis hypogaea* L. (Groundnut) at 50 Days

Methods of nitrogen application	Source nitrogen fertilizer				
	Urea	Ammonium sulphate	Calcium ammonium nitrate	Diammonium phosphate	Mean
T_1 = Full soil	0.50	0.48	0.52	0.54	0.51
T_2 = Full foliar	0.44	0.40	0.42	0.52	0.45
T_3 = 2/3 Soil + 1/3 Foliar	0.48	0.58	0.52	0.56	0.54
T_4 = 1/3 Soil + 2/3 Foliar	0.50	0.46	0.48	0.58	0.51
T_5 = 1/4 Soil + 3/4 Foliar	0.42	0.44	0.46	0.44	0.44
T_6 = 3/4 Soil + 1/4 Foliar	0.52	0.54	0.56	0.52	0.54
Mean	0.48	0.48	0.49	0.53	
	Nitrogen fertilizer (A)		Nitrogen application (B)		A × B
S.Em.±	0.0065		0.0079		0.0159
CD at 5%	0.0185		0.0226		0.0453
Significance	**		**		**

TABLE 1.36 Effect of Different Sources and Methods of Nitrogen Application on Phosphorus (%) of *Arachis hypogaea* L. (Groundnut) at 70 Days

Methods of nitrogen application	Source nitrogen fertilizer				
	Urea	Ammonium sulphate	Calcium ammonium nitrate	Diammonium phosphate	Mean
T_1 = Full soil	0.62	0.58	0.56	0.60	0.59
T_2 = Full foliar	0.54	0.52	0.54	0.59	0.55
T_3 = 2/3 Soil + 1/3 Foliar	0.56	0.66	0.56	0.63	0.60
T_4 = 1/3 Soil + 2/3 Foliar	0.54	0.50	0.54	0.62	0.55
T_5 = 1/4 Soil + 3/4 Foliar	0.49	0.52	0.54	0.52	0.52
T_6 = 3/4 Soil + 1/4 Foliar	0.58	0.64	0.62	0.60	0.61
Mean	0.56	0.57	0.56	0.59	
	Nitrogen fertilizer (A)		Nitrogen application (B)		A × B
S.Em.±	0.0048		0.0059		0.0118
CD at 5%	0.0137		0.0168		0.0337
Significance	**		**		**

TABLE 1.37   Effect of Different Sources and Methods of Nitrogen Application on Phosphorus (%) of *Arachis hypogaea* L. (Groundnut) at 90 Days

Methods of nitrogen application	Source nitrogen fertilizer				
	Urea	Ammonium sulphate	Calcium ammoni- um nitrate	Diam- monium phosphate	Mean
T$_1$ = Full soil	0.68	0.62	0.64	0.66	0.65
T$_2$ = Full foliar	0.58	0.60	0.60	0.68	0.62
T$_3$ = 2/3 Soil + 1/3 Foliar	0.66	0.74	0.62	0.70	0.68
T$_4$ = 1/3 Soil + 2/3 Foliar	0.64	0.60	0.62	0.72	0.65
T$_5$ = 1/4 Soil + 3/4 Foliar	0.58	0.58	0.60	0.58	0.59
T$_6$ = 3/4 Soil + 1/4 Foliar	0.66	0.68	0.70	0.72	0.69
Mean	0.63	0.64	0.63	0.68	
	Nitrogen fertilizer (A)		Nitrogen application (B)		A × B
S.Em.±	0.0057		0.0070		0.0141
CD at 5%	0.0164		0.0201		0.0402
Significance	**		**		**

1.5.1.3.3   Leaf Potassium (%)

Leaf potassium content was significantly affected due to different methods of nitrogen application at 70 and 90 days (Tables 1.38–1.40). At 70 and 90 days, leaf potassium percentages were maximum in 3/4 soil + 1/4 foliar, followed by 2/3 soil + 1/3 foliar method of nitrogen application.

1.5.1.4   Yield Characteristics

Yield characteristics in groundnut at harvest (fresh weight, pod number, pod weight, total pod yield, and oil content were noted to be significant for different methods of nitrogen applications, their sources, and interaction effects (Tables 1.41–1.45). The results in brief have been described in the following subsections.

TABLE 1.38   Effect of Different Sources and Methods of Nitrogen Application on Leaf Potassium (%) of *Arachis hypogaea* L. (Groundnut) at 50 Days

Methods of nitrogen application	Source nitrogen fertilizer				
	Urea	Ammonium sulphate	Calcium ammoni- um nitrate	Diammonium phosphate	Mean
T_1 = Full soil	1.44	1.40	1.42	1.44	1.43
T_2 = Full foliar	1.40	1.38	1.40	1.48	1.42
T_3 = 2/3 Soil + 1/3 Foliar	1.42	1.60	1.52	1.58	1.53
T_4 = 1/3 Soil + 2/3 Foliar	1.44	1.40	1.42	1.58	1.46
T_5 = 1/4 Soil + 3/4 Foliar	1.40	1.42	1.42	1.41	1.41
T_6 = 3/4 Soil + 1/4 Foliar	1.52	1.56	1.56	1.54	1.55
Mean	1.44	1.46	1.46	1.51	

	Nitrogen fertilizer (A)	Nitrogen application (B)	A × B
S.Em.±	0.031	0.039	0.078
CD at 5%	0.091	0.111	0.223
Significance	NS	NS	NS

TABLE 1.39   Effect of Different Sources and Methods of Nitrogen Application on Leaf Potassium (%) of *Arachis hypogaea* L. (Groundnut) at 70 Days

Methods of nitrogen application	Source nitrogen fertilizer				
	Urea	Ammonium sulphate	Calcium ammoni- um nitrate	Diam- monium phosphate	Mean
T_1 = Full soil	1.60	1.62	1.60	1.60	1.61
T_2 = Full foliar	1.50	1.48	1.48	1.64	1.53
T_3 = 2/3 Soil + 1/3 Foliar	1.52	1.82	1.60	1.76	1.68
T_4 = 1/3 Soil + 2/3 Foliar	1.50	1.49	1.52	1.80	1.58
T_5 = 1/4 Soil + 3/4 Foliar	1.48	1.50	1.49	1.46	1.48
T_6 = 3/4 Soil + 1/4 Foliar	1.65	1.78	1.78	1.76	1.74
Mean	1.54	1.62	1.58	1.67	

	Nitrogen fertilizer (A)	Nitrogen application (B)	A × B
S.Em.±	0.037	0.046	0.092
CD at 5%	0.107	0.131	0.263
Significance	NS	**	NS

TABLE 1.40   Effect of Different Sources and Methods of Nitrogen Application on Leaf Potassium (%) of *Arachis hypogaea* L. (Groundnut) at 90 Days

Methods of nitrogen application	Source nitrogen fertilizer				
	Urea	Ammonium sulphate	Calcium ammonium nitrate	Diammonium phosphate	Mean
T_1 = Full soil	1.66	1.60	1.62	1.64	1.63
T_2 = Full foliar	1.56	1.54	1.56	1.70	1.59
T_3 = 2/3 Soil + 1/3 Foliar	1.58	1.98	1.68	1.82	1.77
T_4 = 1/3 Soil + 2/3 Foliar	1.60	1.50	1.54	1.86	1.63
T_5 = 1/4 Soil + 3/4 Foliar	1.52	1.50	1.52	1.48	1.51
T_6 = 3/4 Soil + 1/4 Foliar	1.72	1.82	1.84	1.80	1.80
Mean	1.61	1.66	1.63	1.72	

	Nitrogen fertilizer (A)	Nitrogen application (B)	A × B
S.Em.±	0.038	0.046	0.093
CD at 5%	0.108	0.132	0.265
Significance	NS	**	NS

1.5.1.4.1   Weight/Plant (g)

At harvest, fresh weight of the plants was significantly affected due to different methods of nitrogen application (Table 1.41). The highest value was found in 3/4 soil + 1/4 foliar and was statistically equal to 2/3 soil + 1/3 foliar. Full foliar application of nitrogen application gave the lowest value of fresh weight.

The response of ammonium sulfate was best, statistically equal to diammonium phosphate.

As far as the interaction effect was concerned 2/3 soil + 1/3 foliar application of ammonium sulfate gave the highest value for fresh weight/plant (Table 1.41).

1.5.1.4.2   Number of Pods/Plant

Pod number was significantly affected at harvest due to different methods of nitrogen application, sources, and on their interaction (Table 1.42).

TABLE 1.41 Effect of Different Sources and Methods of Nitrogen Application on Fresh Weight (g/plant) of *Arachis hypogaea* L. (Groundnut) at 50 Days

Methods of nitrogen application	Source nitrogen fertilizer				Mean
	Urea	Ammonium sulphate	Calcium ammonium nitrate	Diammonium phosphate	
T1 = Full soil	322.52	309.10	302.27	317.13	312.76
T2 = Full foliar	313.42	300.62	296.68	341.24	312.99
T3 = 2/3 Soil + 1/3 Foliar	334.60	455.53	338.92	394.52	380.89
T4 = 1/3 Soil + 2/3 Foliar	293.51	283.24	293.61	348.43	304.70
T5 = 1/4 Soil + 3/4 Foliar	286.20	293.23	297.42	310.00	296.71
T6 = 3/4 Soil + 1/4 Foliar	358.21	421.11	428.10	327.62	383.76
Mean	318.08	343.81	326.17	339.82	
	Nitrogen fertilizer (A)		Nitrogen application (B)		A × B
S.Em.±	4.28		5.24		10.48
CD at 5%	12.19		14.93		29.86
Significance	**		**		**

TABLE 1.42 Effect of Different Sources and Methods of Nitrogen Application on Pod Number of *Arachis hypogaea* L. (Groundnut) at Harvest

Methods of nitrogen application	Source nitrogen fertilizer				Mean
	Urea	Ammonium sulphate	Calcium ammonium nitrate	Diammonium phosphate	
T_1 = Full soil	17.34	19.41	19.61	18.63	18.75
T_2 = Full foliar	12.26	10.52	10.02	19.66	13.12
T_3 = 2/3 Soil + 1/3 Foliar	20.67	29.41	19.09	21.33	22.63
T_4 = 1/3 Soil + 2/3 Foliar	20.66	17.34	17.91	20.67	19.15
T_5 = 1/4 Soil + 3/4 Foliar	16.32	16.23	17.21	16.34	16.53
T_6 = 3/4 Soil + 1/4 Foliar	19.34	28.52	25.47	27.00	25.08
Mean	17.77	20.24	18.22	20.61	
	Nitrogen fertilizer (A)		Nitrogen application (B)		A × B
S.Em.±	0.364		0.446		0.892
CD at 5%	1.037		1.270		2.540
Significance	**		**		**

Maximum number of pods were noted in 3/4 soil + 1/4 foliar application. The lowest value was recorded in full foliar application.

The response of diammonium phosphate was best for pod number, statistically equal to ammonium sulfate.

Among the various interactions, 2/3 soil + 1/3 foliar application of ammonium sulfate was found best for pod number, statistically equal to 3/4 soil + 1/4 foliar application of ammonium sulfate as well as 3/4 soil + 1/4 foliar application of diammonium phosphate (Table 1.42).

1.5.1.4.3 Weight of Pod/Plant (g)

A harvest pod weight was significantly affected due to different methods of nitrogen application, their sources, and their interaction (Table 1.43). The highest value was noted for 3/4 soil + 1/4 foliar application, statistically equal to 2/3 soil + 1/3 foliar application.

The response for ammonium sulfate was best, followed by diammonium phosphate for this parameter.

TABLE 1.43 Effect of Different Sources and Methods of Nitrogen Application on Pod Weight (g/plant) of *Arachis hypogaea* L. (Groundnut) at Harvest

Methods of nitrogen application	Source nitrogen fertilizer				
	Urea	Ammonium sulphate	Calcium ammonium nitrate	Diam-monium phosphate	Mean
T$_1$ = Full soil	13.89	13.40	13.40	14.00	13.67
T$_2$ = Full foliar	10.60	10.60	11.20	16.40	12.20
T$_3$ = 2/3 Soil + 1/3 Foliar	14.20	23.40	13.40	14.90	16.48
T$_4$ = 1/3 Soil + 2/3 Foliar	11.90	12.20	12.20	14.80	12.78
T$_5$ = 1/4 Soil + 3/4 Foliar	11.40	12.40	12.60	10.40	11.70
T$_6$ = 3/4 Soil + 1/4 Foliar	13.60	19.10	19.40	15.80	16.98
Mean	12.60	15.18	13.70	14.38	
	Nitrogen fertilizer (A)		Nitrogen application (B)		A × B
S.Em.±	0.208		0.255		0.510
CD at 5%	0.593		0.727		1.454
Significance	**		**		**

Regarding interactions, 2/3 soil + 1/3 foliar application of ammonium sulfate gave the highest value (Table 1.43).

1.5.1.4.4 Pod Yield (kg/ha)

At harvest, total pod yield was significantly affected due to different methods of nitrogen application, sources, and their interactions (Table 1.44). The highest value was noted in 3/4 soil + 1/4 foliar, followed by 2/3 soil + 1/3 foliar; the value differed critically with each other. This was followed by full soil application. Full foliar application proved to yield the lowest value.

The response of diammonium phosphate was best, followed by ammonium sulfate for total pod yield. It was interesting to note that 3/4 soil + 1/4 foliar gave 8.5% more pod yield as compared to full soil application method (Table 1.44).

As far as the interaction effect was concerned, 2/3 soil + 1/3 foliar application of ammonium sulfate was best for pod yield, statistically equal

TABLE 1.44   Effect of Different Sources and Methods of Nitrogen Application on Pod Yield (kg/ha) of *Arachis hypogaea* L. (Groundnut) at Harvest

Methods of nitrogen application	Source nitrogen fertilizer				
	Urea	Am-monium sulphate	Calcium ammonium nitrate	Diam-monium phosphate	Mean
T_1 = Full soil	1570.00	1693.00	1620.00	1600.00	1620.75
T_2 = Full foliar	1386.00	1008.00	1092.00	1728.00	1303.50
T_3 = 2/3 Soil + 1/3 Foliar	1408.00	1918.00	1612.00	1840.00	1694.50
T_4 = 1/3 Soil + 2/3 Foliar	1460.00	1260.00	1480.00	1864.00	1516.00
T_5 = 1/4 Soil + 3/4 Foliar	1188.00	1296.00	1248.00	1134.00	1216.50
T_6 = 3/4 Soil + 1/4 Foliar	1756.00	1840.00	1816.00	1626.00	1759.50
Mean	1461.33	1502.50	1478.00	1632.00	
	Nitrogen fertilizer (A)		Nitrogen application (B)		A × B
S.Em.±	17.55		21.50		43.00
CD at 5%	49.97		61.20		122.41
Significance	**		**		**

to 1/3 soil + 2/3 foliar application of diammonium phosphate as well as 2/3 soil + 1/3 foliar application of diammonium phosphate (Table 1.44).

1.5.1.5   Total Oil Content (%)

Total oil content percentage in seeds at harvest in groundnut was significantly affected due to different methods of nitrogen application, sources, and their interaction (Table 1.45). Total seed oil percentage was maximum for full soil application, statistically equal to 3/4 soil + 1/4 foliar as well as 2/3 soil + 1/3 foliar method of nitrogen application. The lowest value was found with full foliar application.

The response of ammonium sulfate was best for this important attribute in groundnut.

As far as the interaction effect was concerned, highest value was obtained by 2/3 soil + 1/3 foliar application of ammonium sulfate for this yield attributing character.

TABLE 1.45   Effect of Different Sources and Methods of Nitrogen Application on Total Seed Oil (%) of *Arachis hypogaea* L. (Groundnut) at Harvest

Methods of nitrogen application	Source nitrogen fertilizer				
	Urea	Ammonium sulphate	Calcium ammonium nitrate	Diammonium phosphate	Mean
T_1 = Full soil	45.90	45.80	46.00	46.20	45.98
T_2 = Full foliar	46.00	44.00	44.00	45.00	44.75
T_3 = 2/3 Soil + 1/3 Foliar	45.00	48.00	44.00	44.00	45.25
T_4 = 1/3 Soil + 2/3 Foliar	46.20	45.00	45.00	46.00	45.55
T_5 = 1/4 Soil + 3/4 Foliar	45.00	45.90	45.00	45.10	45.25
T_6 = 3/4 Soil + 1/4 Foliar	46.00	46.09	46.08	44.09	45.57
Mean	45.68	45.80	45.01	45.07	
	Nitrogen fertilizer (A)		Nitrogen application (B)		A × B
S.Em.±	0.200		0.245		0.491
CD at 5%	0.570		0.699		1.398
Significance	*		*		**

1.5.2   EXPERIMENT 2

The second field experiment was conducted in 2010 on *Arachis hypo-gaea* L. (groundnut) var. TG 37-A during the kharif (summer) season to determine the effect of agrotechniques, different seed rates (row to plant distances), and dates of sowing (at one month intervals), and their interaction in a factorial randomized design. The crop was assessed by growth characteristics; leaf chlorophylls and carotenoid contents; leaf nutrient (NPK) contents; sampled at pre-flowering (50 days), flowering (70 days), and post-flowering (90 days) stages as well as at harvest yield characteristics, pod yield, and oil yield (%). The data are summarized in Tables 1.46–1.84.

1.5.2.1   Growth Characteristics

The growth characteristics consisted of germination% at 15 days after sowing, and three times (50, 70, and 90 days) samplings for fresh weight, leaf number, plant height, root length, dry weight, leaf chlorophyll a, leaf chlorophyll b, and leaf carotenoid content. The results in brief have been described in the following subsections.

1.5.2.1.1   Germination %

Germination % was significantly affected by changing dates of sowing groundnut. The highest value was noted in May 5 sowing and lowest in September 5 (Table 1.46).

1.5.2.1.2   Fresh Weight (g/plant)

Fresh weight of the plant was significantly affected by changing dates of sowing, row to plant distances, and interactions at the three growth stages (Tables 1.47–1.49). At 50 days, the effect of May 5 sowing was best, statistically equal to June 5 sowing. The lowest value was noted in September 5. At 70 and 90 days, again May 5 sowing was best, similar to at 50 days (Tables 1.47–1.49).

TABLE 1.46   Effect of Seed Rate and Dates of Sowing on Germination (%) of *Arachis hypogaea* L. (Groundnut)

Seed rate/plant density	Date of sowing					
	5 May	5 June	5 July	5 August	5 September	Mean
T$_1$=40 x 10 cm	86.67	80.00	68.00	60.00	60.00	67.00
T$_2$=40 x 5 cm	86.65	79.34	66.67	60.00	53.34	64.84
T$_3$=30 x 10 cm	86.34	79.33	66.66	60.00	56.67	65.67
T$_4$=30 x 5 cm	86.66	80.00	66.67	63.34	56.00	66.50
T$_5$=20 x 10 cm	86.67	79.67	70.00	70.00	53.34	68.25
T$_6$=20 x 5 cm	86.67	80.34	66.34	60.00	53.33	65.00
Mean	86.61	79.78	67.39	62.22	55.45	

	Date of sowing (A)	Seed rate/Plant density (B)	A × B
S.Em.±	1.03	1.13	2.53
CD at 5%	2.93	3.21	7.17
Significance	**	NS	NS

TABLE 1.47   Effect of Seed Rate and Dates of Sowing on Fresh Weight (g/plant) of *Arachis hypogaea* L. (Groundnut) at 50 Days

Seed rate/ plant density	Date of sowing					
	5 May	5 June	5 July	5 August	5 September	Mean
T$_1$=40 x 10 cm	148.34	142.67	138.34	106.67	91.00	119.67
T$_2$=40 x 5 cm	135.33	126.34	125.00	108.66	81.00	110.25
T$_3$=30 x 10 cm	143.00	147.66	138.00	107.67	78.34	117.92
T$_4$=30 x 5 cm	138.97	141.00	132.00	103.67	76.33	113.25
T$_5$=20 x 10 cm	140.34	142.34	123.00	101.65	82.00	112.25
T$_6$=20 x 5 cm	122.00	116.33	112.34	98.00	74.67	100.34
Mean	138.00	136.06	128.11	104.39	80.56	

	Date of sowing (A)	Seed rate/Plant density (B)	A × B
S.Em.±	3.52	3.85	8.62
CD at 5%	9.96	10.92	24.41
Significance	**	**	NS

TABLE 1.48 Effect of Seed Rate and Dates of Sowing on Fresh Weight (g/plant) of *Arachis hypogaea* L. (Groundnut) at 70 Days

Seed rate/plant density	Date of sowing					
	5 May	5 June	5 July	5 August	5 September	Mean
T_1 = 40 x 10 cm	296.00	247.34	210.00	191.12	130.00	194.62
T_2 = 40 x 5 cm	225.34	184.32	157.00	124.00	101.00	141.58
T_3 = 30 x 10 cm	232.30	185.67	139.34	136.00	121.67	145.67
T_4 = 30 x 5 cm	209.68	178.00	119.00	113.68	94.34	126.26
T_5 = 20 x 10 cm	126.66	129.66	124.35	129.65	110.68	123.59
T_6 = 20 x 5 cm	131.00	103.00	127.33	102.34	86.32	104.75
Mean	203.50	171.33	146.17	132.80	107.34	

	Date of sowing (A)	Seed rate/Plant density (B)	A × B
S.Em.±	3.32	3.64	8.15
CD at 5%	9.42	10.32	23.07
Significance	**	**	**

TABLE 1.49 Effect of Seed Rate and Dates of Sowing on Fresh Weight (g/plant) of *Arachis hypogaea* L. (Groundnut) at 90 Days

Seed rate/plant density	Date of sowing					
	5 May	5 June	5 July	5 August	5 September	Mean
T_1 = 40 x 10 cm	306.67	249.32	214.67	222.33	199.34	221.42
T_2 = 40 x 5 cm	292.34	210.00	208.00	194.66	183.35	199.00
T_3 = 30 x 10 cm	269.00	240.00	222.00	210.00	165.28	209.32
T_4 = 30 x 5 cm	280.65	211.00	208.00	178.00	171.67	192.17
T_5 = 20 x 10 cm	197.00	216.67	212.00	202.00	144.65	193.83
T_6 = 20 x 5 cm	178.67	178.67	173.35	155.00	111.34	154.59
Mean	254.06	217.61	206.34	193.67	162.61	

	Date of sowing (A)	Seed rate/Plant density (B)	A × B
S.Em.±	3.31	3.63	8.12
CD at 5%	9.38	10.28	22.99
Significance	**	**	**

The response to 40 × 10 cm row to plant spacing was found best at almost all growth stages; the lowest value was noted for 20 × 5 cm spacing.

As far the interaction effect was concerned, the best value was noted for May 5 sowing with 40 × 10 cm spacing followed by May 5 sowing with 40 × 5 cm spacing at both 70 and 90 days.

1.5.2.1.3   Leaf Number

Leaf number was significantly affected by changing dates of sowing and population density at all growth stages (Tables 1.50–1.52). At all the growth stages, the highest value was noted in May 5 sowing and the lowest in September 5. The response to 40 × 10 cm spacing was best followed by 40 × 5 cm spacing at all growth stages.

The interaction effect at 70 days was best in May 5 sowing with 40 × 10 cm spacing, statistically equal to May 5 sowing with 40 × 10 cm spacing. At 90 days also, a similar situation was noted (Tables 1.50–1.52).

1.5.2.1.4   Plant Height (cm)

Plant height in groundnut was significantly affected due to variations in sowing dates, population density, and their interaction at all the growth stages studied (Tables 1.53–1.55). The effect of May 5 sowing was best and lowest in September 5 at all growth stages. The response of 40 × 10 cm population density was best followed by 40 × 10 cm and 30 × 10 cm population density, respectively, at all the growth stages (Tables 1.53–1.55).

As far as the interaction effect was concerned, May 5 sowing with 40 × 10 cm spacing was best followed by May 5 sowing with 40 × 5 cm spacing as well as May 5 sowing with 30 × 10 cm spacing, respectively (Tables 1.53–1.55).

1.5.2.1.5   Root Length (cm)

Root length in groundnut was significantly affected due to change in date of sowing, population density, and their interaction, except at 90 days, for all growth stages (Tables 1.56–1.58). At all the three growth stages, May

TABLE 1.50 Effect of Seed Rate and Dates of Sowing Leaf Number/Plant of *Arachis hypogaea* L. (Groundnut) at 50 Days

Seed rate/plant density	Date of sowing					
	5 May	5 June	5 July	5 August	5 September	Mean
T_1=40 x 10 cm	43.00	37.34	34.67	31.34	26.66	32.50
T_2=40 x 5 cm	38.67	35.00	26.65	26.35	24.90	28.23
T_3=30 x 10 cm	38.00	35.00	25.00	28.00	25.34	28.34
T_4=30 x 5 cm	33.66	30.33	20.66	21.67	22.68	23.84
T_5=20 x 10 cm	30.34	26.32	21.68	24.00	20.67	23.17
T_6=20 x 5 cm	22.33	23.00	18.00	20.68	18.00	19.92
Mean	34.33	31.17	24.44	25.34	23.04	

	Date of sowing (A)	Seed rate/Plant density (B)	A × B
S.Em.±	0.68	0.74	1.66
CD at 5%	1.92	2.10	4.71
Significance	**	**	NS

TABLE 1.51 Effect of Seed Rate and Dates of Sowing on Leaf Number/Plant of *Arachis hypogaea* L. (Groundnut) at 70 Days

Seed rate/plant density	Date of sowing					
	5 May	5 June	5 July	5 August	5 September	Mean
T_1=40 x 10 cm	69.90	54.00	44.00	39.67	28.34	41.50
T_2=40 x 5 cm	63.69	54.67	36.66	34.00	23.68	37.25
T_3=30 x 10 cm	55.65	53.34	33.33	31.00	26.67	36.09
T_4=30 x 5 cm	49.00	44.68	29.67	21.90	20.65	29.23
T_5=20 x 10 cm	45.35	40.35	22.34	24.68	23.66	27.76
T_6=20 x 5 cm	43.00	40.66	21.00	23.00	17.34	25.50
Mean	54.43	47.95	31.17	29.04	23.39	

	Date of sowing (A)	Seed rate/Plant density (B)	A × B
S.Em.±	0.76	0.83	1.87
CD at 5%	2.16	2.37	5.30
Significance	**	**	**

TABLE 1.52   Effect of Seed Rate and Dates of Sowing on Leaf Number/Plant of *Arachis hypogaea* L. (Groundnut) at 90 Days

Seed rate/plant density	Date of sowing					
	5 May	5 June	5 July	5 August	5 September	Mean
T_1 = 40 x 10 cm	66.35	53.62	45.34	36.90	21.67	39.38
T_2 = 40 x 5 cm	61.38	53.00	34.00	36.80	23.34	36.79
T_3 = 30 x 10 cm	53.67	54.36	36.32	34.90	24.00	37.40
T_4 = 30 x 5 cm	55.35	51.34	27.35	35.00	25.35	34.76
T_5 = 20 x 10 cm	44.67	51.00	27.00	28.00	24.00	32.50
T_6 = 20 x 5 cm	41.68	42.00	24.38	24.67	21.66	28.18
Mean	53.85	50.89	32.40	32.71	23.34	

	Date of sowing (A)	Seed rate/Plant density (B)	A × B
S.Em.±	0.90	0.99	2.22
CD at 5%	2.57	2.81	6.30
Significance	**	**	**

TABLE 1.53   Effect of Seed Rate and Dates of Sowing on Plant Height (cm) of *Arachis hypogaea* L. (Groundnut) at 50 Days

Seed rate/plant density	Date of sowing					
	5 May	5 June	5 July	5 August	5 September	Mean
T_1 = 40 x 10 cm	48.17	43.32	46.34	38.00	30.00	39.42
T_2 = 40 x 5 cm	46.50	43.34	45.00	35.34	31.67	38.84
T_3 = 30 x 10 cm	45.50	40.68	43.00	33.67	30.00	36.84
T_4 = 30 x 5 cm	44.67	40.35	40.32	35.32	29.90	36.47
T_5 = 20 x 10 cm	44.84	38.32	37.00	35.00	27.68	34.50
T_6 = 20 x 5 cm	44.85	39.33	32.68	31.34	22.00	31.34
Mean	45.76	40.89	40.72	34.78	28.54	

	Date of sowing (A)	Seed rate/Plant density (B)	A × B
S.Em.±	0.60	0.66	1.47
CD at 5%	1.70	1.87	4.18
Significance	**	**	*

TABLE 1.54 Effect of Seed Rate and Dates of Sowing on Plant Height (cm) of *Arachis hypogaea* L. (Groundnut) at 70 Days

Seed rate/plant density	Date of sowing					
	5 May	5 June	5 July	5 August	5 September	Mean
T_1 = 40 x 10 cm	76.00	71.67	68.34	44.35	47.69	58.01
T_2 = 40 x 5 cm	74.67	65.69	66.00	43.30	36.67	52.92
T_3 = 30 x 10 cm	74.00	55.32	52.00	47.00	31.00	46.33
T_4 = 30 x 5 cm	76.68	52.94	46.90	46.31	38.66	46.20
T_5 = 20 x 10 cm	67.34	47.67	50.66	41.32	35.00	43.66
T_6 = 20 x 5 cm	65.35	48.32	43.34	38.68	31.00	40.34
Mean	72.34	56.94	54.54	43.49	36.67	

	Date of sowing (A)	Seed rate/Plant density (B)	A × B
S.Em.±	0.88	0.96	2.15
CD at 5%	2.49	2.73	6.11
Significance	**	**	**

TABLE 1.55 Effect of Seed Rate and Dates of sowing on Plant Height (cm) of *Arachis hypogaea* L. (Groundnut) at 90 Days

Seed rate/plant density	Date of sowing					
	5 May	5 June	5 July	5 August	5 September	Mean
T_1 = 40 x 10 cm	76.90	72.00	71.00	45.00	48.00	59.00
T_2 = 40 x 5 cm	74.90	65.67	68.00	44.66	36.90	53.81
T_3 = 30 x 10 cm	77.67	56.65	53.35	44.00	29.36	45.84
T_4 = 30 x 5 cm	77.68	54.00	47.67	51.00	29.33	45.50
T_5 = 20 x 10 cm	69.34	48.00	49.33	46.34	38.35	45.51
T_6 = 20 x 5 cm	67.65	49.34	46.35	39.90	35.00	42.65
Mean	74.02	57.61	55.95	45.15	36.16	

	Date of sowing (A)	Seed rate/Plant density (B)	A × B
S.Em.±	0.82	0.90	2.01
CD at 5%	2.33	2.55	5.71
Significance	**	**	**

5 sowing gave highest value; the lowest value was noted in September 5 sowing. Similarly, the response of 40 × 10 cm population density was best at all the growth stages followed by 40 × 5 cm (Tables 1.56–1.58).

The interaction effect was noted best in May 5 sowing with 40 × 10 cm spacing at 50 and 70 days (Tables 1.56–1.58); also in June 5 sowing with 40 × 10 cm spacing interaction.

1.5.2.1.6   Dry Weight (g/plant)

Dry weight of the groundnut plant was significantly affected due to variations in date of sowing, population density, and their interaction at all growth stages studied (Tables 1.59–1.61). At all the growth stages, the effect of ,May 5 sowing gave maximum value; minimum value was given by September 5 sowing (Tables 1.59–1.61). The response to 40 ×10 cm and 40 × 5 cm population densities were noted better than other population densities at all growth stages.

As far the interaction effect was concerned, highest value was given by May 5 sowing with 40 × 10 cm spacing; also June 5 spacing with 40 × 10 cm spacing at 50 days, 70 days, and 90 days (Tables 1.59–1.61).

TABLE 1.56   Effect of Seed Rate and Dates of Sowing on Root Length (cm) of *Arachis hypogaea* L. (Groundnut) at 50 Days

Seed rate/plant density	Date of sowing					
	5 May	5 June	5 July	5 August	5 September	Mean
T_1 = 40 x 10 cm	7.90	8.00	7.08	7.01	7.09	7.30
T_2 = 40 x 5 cm	7.80	7.90	7.10	7.05	7.01	7.27
T_3 = 30 x 10 cm	7.40	7.10	7.06	7.01	7.02	7.05
T_4 = 30 x 5 cm	7.45	6.89	7.05	7.10	6.99	7.01
T_5 = 20 x 10 cm	7.10	7.10	7.01	7.00	6.99	7.03
T_6 = 20 x 5 cm	7.05	6.99	6.79	6.80	6.39	6.74
Mean	7.45	7.33	7.02	7.00	6.92	

	Date of sowing (A)	Seed rate/Plant density (B)	A × B
S.Em.±	0.066	0.072	0.162
CD at 5%	0.187	0.205	0.460
Significance	**	**	*

TABLE 1.57 Effect of Seed Rate and Dates of Sowing on Root Length (cm) of *Arachis hypogaea* L. (Groundnut) at 70 Days

Seed rate/ plant density	Date of sowing					
	5 May	5 June	5 July	5 August	5 September	Mean
T₁=40 x 10 cm	9.29	8.90	8.05	7.20	7.10	7.81
T₂=40 x 5 cm	9.10	8.40	8.02	7.70	7.08	7.80
T₃=30 x 10 cm	9.19	8.10	8.10	7.80	7.10	7.78
T₄=30 x 5 cm	9.20	8.50	8.11	7.20	7.00	7.70
T₅=20 x 10 cm	8.80	8.40	8.70	7.10	7.01	7.80
T₆=20 x 5 cm	8.30	8.41	8.50	7.50	7.00	7.85
Mean	8.98	8.45	8.25	7.42	7.05	

	Date of sowing (A)	Seed rate/Plant density (B)	A × B
S.Em.±	0.030	0.033	0.073
CD at 5%	0.085	0.093	0.209
Significance	**	*	**

TABLE 1.58 Effect of Seed Rate and Dates of Sowing on Root Length (cm) of *Arachis hypogaea* L. (Groundnut) at 90 Days

Seed rate/plant density	Date of sowing					
	5 May	5 June	5 July	5 August	5 September	Mean
T₁=40 x 10 cm	12.67	9.59	9.00	9.60	7.78	8.99
T₂=40 x 5 cm	12.17	10.50	9.17	9.60	7.70	9.24
T₃=30 x 10 cm	11.50	9.84	9.34	9.57	7.20	8.99
T₄=30 x 5 cm	12.00	9.80	9.67	9.47	7.34	9.07
T₅=20 x 10 cm	11.84	9.34	8.84	9.34	7.27	8.70
T₆=20 x 5 cm	11.00	9.50	8.97	8.84	7.67	8.75
Mean	11.86	9.76	9.17	9.40	7.49	

	Date of sowing (A)	Seed rate/Plant density (B)	A × B
S.Em.±	0.104	0.114	0.256
CD at 5%	0.295	0.324	0.724
Significance	**	**	NS

TABLE 1.59   Effect of Seed Rate and Dates of Sowing on Dry Weight (g/plant) of *Arachis hypogaea* L. (Groundnut) at 50 Days

Seed rate/ plant density	Date of sowing					
	5 May	5 June	5 July	5 August	5 September	Mean
T_1 = 40 x 10 cm	11.54	12.67	10.34	9.34	9.20	10.39
T_2 = 40 x 5 cm	9.77	9.27	9.14	8.80	8.20	8.85
T_3 = 30 x 10 cm	9.70	9.57	9.34	8.54	8.60	9.01
T_4 = 30 x 5 cm	8.99	8.50	8.47	8.74	8.20	8.48
T_5 = 20 x 10 cm	8.94	8.50	8.67	8.14	8.37	8.42
T_6 = 20 x 5 cm	8.40	7.54	6.94	8.54	8.87	7.97
Mean	9.56	9.34	8.82	8.68	8.57	

	Date of sowing (A)	Seed rate/Plant density (B)	A × B
S.Em.±	0.072	0.078	0.176
CD at 5%	0.203	0.223	0.499
Significance	**	**	**

TABLE 1.60   Effect of Seed Rate and Dates of Sowing on Dry Weight (g/plant) of *Arachis hypogaea* L. (Groundnut) at 70 Days

Seed rate/plant density	Date of sowing					
	5 May	5 June	5 July	5 August	5 September	Mean
T_1 = 40 x 10 cm	29.07	26.00	20.14	18.20	16.00	20.09
T_2 = 40 x 5 cm	25.14	27.00	25.34	19.67	15.20	21.80
T_3 = 30 x 10 cm	26.07	27.14	24.67	17.20	14.34	20.84
T_4 = 30 x 5 cm	25.80	25.34	23.87	20.74	10.87	20.21
T_5 = 20 x 10 cm	24.20	23.94	22.80	20.87	12.14	19.94
T_6 = 20 x 5 cm	20.90	20.60	20.47	18.67	12.20	17.99
Mean	25.20	25.00	22.88	19.23	13.46	

	Date of sowing (A)	Seed rate/Plant density (B)	A × B
S.Em.±	0.457	0.501	1.120
CD at 5%	1.295	1.418	3.172
Significance	**	**	**

TABLE 1.61 Effect of Seed Rate and Dates of Sowing on Dry Weight (g/plant) of *Arachis hypogaea* L. (Groundnut) at 90 Days

Seed rate/ plant density	Date of sowing					
	5 May	5 June	5 July	5 August	5 September	Mean
T_1=40 x 10 cm	61.34	49.34	46.87	44.47	31.80	43.12
T_2=40 x 5 cm	58.47	48.07	44.54	38.34	26.64	39.40
T_3=30 x 10 cm	59.00	48.00	44.34	37.80	21.00	37.79
T_4=30 x 5 cm	56.14	50.94	40.00	36.00	24.17	37.78
T_5=20 x 10 cm	49.40	50.34	38.47	34.67	22.00	36.37
T_6=20 x 5 cm	49.14	45.74	34.60	31.00	20.20	32.89
Mean	55.58	48.74	41.47	37.05	24.30	

	Date of sowing (A)	Seed rate/Plant density (B)	A × B
S.Em.±	0.733	0.803	1.796
CD at 5%	2.076	2.274	5.086
Significance	**	**	*

1.5.2.1.7 Leaf Chlorophyll a (g/g)

Leaf chlorophyll a content was significantly affected due to change in the date of sowing, row to plant distances, and their interaction at all growth stages (Tables 1.62–1.64). At all stages, the effect of May 5 sowing was best; the lowest value was given by September 5 sowing. The response to 40 × 10 cm population density was noted best at all growth stages, followed by 40 × 5 cm and then 30 × 10 cm, respectively.

As far as the interaction effect was concerned, May 5 sowing with 40 × 10 cm spacing and May 5 sowing with 40 × 5 cm spacing showed superiority over others at all growth stages (Tables 1.62–1.64).

1.5.2.1.8 Leaf Chlorophyll b (mg/g)

Leaf chlorophyll b was significantly affected due to variations in the date of sowing, population density, and their interaction at all the three growth stages studied (Tables 1.65–1.67). The effect of May 5 sowing for this

TABLE 1.62 Effect of Seed Rate and Dates of Sowing on Chlorophyll 'a' (mg/g) of *Arachis hypogaea* L. (groundnut) at 50 Days

Seed rate/plant density	Date of sowing					
	5 May	**5 June**	**5 July**	**5 August**	**5 September**	**Mean**
T_1=40 x 10 cm	0.76	0.66	0.64	0.58	0.56	0.61
T_2=40 x 5 cm	0.77	0.66	0.57	0.53	0.49	0.56
T_3=30 x 10 cm	0.73	0.64	0.58	0.55	0.47	0.56
T_4=30 x 5 cm	0.75	0.66	0.54	0.47	0.47	0.54
T_5=20 x 10 cm	0.66	0.64	0.46	0.46	0.45	0.50
T_6=20 x 5 cm	0.67	0.62	0.47	0.47	0.46	0.51
Mean	0.72	0.65	0.54	0.51	0.48	

	Date of sowing (A)	Seed rate/Plant density (B)	A × B
S.Em.±	0.0062	0.0068	0.0154
CD at 5%	0.0178	0.0195	0.0436
Significance	**	**	**

TABLE 1.63 Effect of Seed Rate and Dates of Sowing on Chlorophyll 'a' (mg/g) of *Arachis hypogaea* L. (Groundnut) at 70 Days

Seed rate/ plant density	Date of sowing					
	5 May	**5 June**	**5 July**	**5 August**	**5 September**	**Mean**
T_1=40 x 10 cm	0.78	0.64	0.57	0.56	0.50	0.57
T_2=40 x 5 cm	0.74	0.65	0.57	0.51	0.47	0.55
T_3=30 x 10 cm	0.74	0.60	0.56	0.53	0.39	0.52
T_4=30 x 5 cm	0.73	0.63	0.55	0.55	0.41	0.54
T_5=20 x 10 cm	0.64	0.63	0.49	0.45	0.38	0.49
T_6=20 x 5 cm	0.68	0.59	0.50	0.40	0.38	0.47
Mean	0.72	0.62	0.54	0.50	0.42	

	Date of sowing (A)	Seed rate/Plant density (B)	A × B
S.Em.±	0.0072	0.0079	0.0177
CD at 5%	0.0205	0.0224	0.0502
Significance	**	**	**

TABLE 1.64 Effect of Seed Rate and Dates of Sowing on Chlorophyll 'a' (mg/g) of *Arachis hypogaea* L. (Groundnut) at 90 Days

Seed rate/plant density	Date of sowing					
	5 May	5 June	5 July	5 August	5 September	Mean
T_1 = 40 x 10 cm	0.78	0.64	0.57	0.56	0.50	0.57
T_2 = 40 x 5 cm	0.74	0.65	0.57	0.51	0.47	0.55
T_3 = 30 x 10 cm	0.74	0.60	0.56	0.53	0.39	0.52
T_4 = 30 x 5 cm	0.73	0.63	0.55	0.55	0.41	0.54
T_5 = 20 x 10 cm	0.64	0.63	0.49	0.45	0.38	0.49
T_6 = 20 x 5 cm	0.68	0.59	0.50	0.40	0.38	0.47
Mean	0.72	0.62	0.54	0.50	0.42	

	Date of sowing (A)	Seed rate/Plant density (B)	A × B
S.Em.±	0.0039	0.0043	0.0096
CD at 5%	0.0111	0.0121	0.0272
Significance	**	**	**

TABLE 1.65 Effect of Seed Rate and Dates of Sowing on chlorophyll 'b' (mg/g) of *Arachis hypogaea* L. (Groundnut) at 50 Days

Seed rate/plant density	Date of sowing					
	5 May	5 June	5 July	5 August	5 September	Mean
T_1 = 40 x 10 cm	0.47	0.45	0.44	0.41	0.43	0.43
T_2 = 40 x 5 cm	0.46	0.45	0.40	0.38	0.35	0.40
T_3 = 30 x 10 cm	0.54	0.46	0.42	0.35	0.35	0.40
T_4 = 30 x 5 cm	0.58	0.47	0.44	0.35	0.36	0.41
T_5 = 20 x 10 cm	0.43	0.44	0.42	0.38	0.36	0.40
T_6 = 20 x 5 cm	0.46	0.44	0.42	0.32	0.36	0.39
Mean	0.49	0.45	0.42	0.37	0.37	

	Date of sowing (A)	Seed rate/Plant density (B)	A × B
S.Em.±	0.0051	0.0056	0.0126
CD at 5%	0.0146	0.0160	0.0358
Significance	**	**	**

TABLE 1.66   Effect of Seed Rate and Dates of Sowing on Chlorophyll 'b' (mg/g) of *Arachis hypogaea* L. (Groundnut) at 70 Days

Seed rate/plant density	Date of sowing					
	5 May	5 June	5 July	5 August	5 September	Mean
T$_1$=40 x 10 cm	0.48	0.48	0.45	0.35	0.32	0.40
T$_2$=40 x 5 cm	0.45	0.45	0.44	0.36	0.31	0.39
T$_3$=30 x 10 cm	0.55	0.44	0.45	0.39	0.31	0.40
T$_4$=30 x 5 cm	0.57	0.45	0.49	0.39	0.31	0.41
T$_5$=20 x 10 cm	0.56	0.43	0.46	0.35	0.26	0.38
T$_6$=20 x 5 cm	0.53	0.44	0.42	0.34	0.26	0.37
Mean	0.52	0.45	0.45	0.36	0.30	

	Date of sowing (A)	Seed rate/Plant density (B)	A × B
S.Em.±	0.0060	0.0066	0.0148
CD at 5%	0.0172	0.0188	0.0421
Significance	**	**	**

TABLE 1.67   Effect of Seed Rate and Dates of Sowing on chlorophyll 'b' (mg/g) of *Arachis hypogaea* L. (Groundnut) at 90 Days

Seed rate/plant density	Date of sowing					
	5 May	5 June	5 July	5 August	5 September	Mean
T$_1$=40 x 10 cm	0.42	0.43	0.34	0.23	0.16	0.29
T$_2$=40 x 5 cm	0.41	0.39	0.34	0.24	0.16	0.28
T$_3$=30 x 10 cm	0.47	0.39	0.31	0.25	0.13	0.27
T$_4$=30 x 5 cm	0.48	0.33	0.30	0.26	0.13	0.26
T$_5$=20 x 10 cm	0.40	0.31	0.33	0.18	0.14	0.24
T$_6$=20 x 5 cm	0.43	0.31	0.29	0.11	0.12	0.21
Mean	0.44	0.36	0.32	0.21	0.14	

	Date of sowing (A)	Seed rate/Plant density (B)	A × B
S.Em.±	0.0054	0.0059	0.0133
CD at 5%	0.0154	0.0168	0.0377
Significance	**	**	**

parameter was noted maximum at all the stages, while the lowest value was found in September 5 sowing. The response to 40 × 10 cm spacing was found almost highest at all the growth stages, followed by 30 × 5 cm and 30 × 10 cm spacing, respectively (Tables 1.66–1.67).

The interaction effect was also significantly best at 50 days in May 5 sowing with 40 × 10 cm spacing, followed by May 5 sowing with 30 × 10 cm spacing and May 5 sowing with 30 × 5 cm spacing. At 70 and 90 days, May 5 sowing with 30 × 5 cm spacing and May 5 sowing with 30 × 10 cm spacing proved superior over other combinations of interaction (Tables 1.65–1.67).

1.5.2.1.9   Leaf Carotenoid (mg/g)

Leaf carotenoid content was significantly affected due to changed date of sowing, population density, and their interaction at all the three growth stages (Tables 1.68–1.70). The effect of May 5 sowing was maximum at all the growth stages; the minimum effect was recorded in September 5

TABLE 1.68   Effect of Seed Rate and Dates of Sowing on Carotenoid (mg/g) of *Arachis hypogaea* L. (Groundnut) at 50 Days

Seed rate/plant density	Date of sowing					Mean
	5 May	5 June	5 July	5 August	5 September	
T$_1$ = 40 x 10 cm	0.16	0.15	0.11	0.13	0.08	0.12
T$_2$ = 40 x 5 cm	0.18	0.17	0.11	0.14	0.09	0.13
T$_3$ = 30 x 10 cm	0.19	0.14	0.12	0.13	0.08	0.12
T$_4$ = 30 x 5 cm	0.18	0.15	0.14	0.10	0.08	0.12
T$_5$ = 20 x 10 cm	0.17	0.14	0.10	0.10	0.08	0.11
T$_6$ = 20 x 5 cm	0.15	0.12	0.11	0.11	0.08	0.11
Mean	0.17	0.15	0.12	0.12	0.08	

	Date of sowing (A)	Seed rate/Plant density (B)	A × B
S.Em.±	0.0024	0.0027	0.0060
CD at 5%	0.0070	0.0076	0.0172
Significance	**	**	**

TABLE 1.69 Effect of Seed Rate and Dates of Sowing on Carotenoid (mg/g) of *Arachis hypogaea* L. (Groundnut) at 70 Days

Seed rate/ plant density	Date of sowing					Mean
	5 May	5 June	5 July	5 August	5 September	
T_1=40 x 10 cm	0.19	0.18	0.17	0.11	0.04	0.13
T_2=40 x 5 cm	0.19	0.14	0.18	0.11	0.04	0.12
T_3=30 x 10 cm	0.20	0.16	0.16	0.09	0.08	0.12
T_4=30 x 5 cm	0.21	0.17	0.15	0.09	0.07	0.12
T_5=20 x 10 cm	0.10	0.14	0.10	0.07	0.05	0.09
T_6=20 x 5 cm	0.19	0.14	0.10	0.08	0.03	0.09
Mean	0.18	0.16	0.14	0.09	0.05	

	Date of sowing (A)	Seed rate/Plant density (B)	A × B
S.Em.±	0.0030	0.0033	0.0075
CD at 5%	0.0086	0.0094	0.0212
Significance	**	**	**

TABLE 1.70 Effect of Seed Rate and Dates of Sowing on Carotenoid (mg/g) of *Arachis hypogaea* L. (Groundnut) at 90 Days

Seed rate/plant density	Date of sowing					Mean
	5 May	5 June	5 July	5 August	5 September	
T_1=40 x 10 cm	0.18	0.16	0.16	0.08	0.08	0.12
T_2=40 x 5 cm	0.13	0.16	0.16	0.07	0.05	0.11
T_3=30 x 10 cm	0.18	0.14	0.17	0.06	0.07	0.11
T_4=30 x 5 cm	0.17	0.13	0.06	0.09	0.09	0.09
T_5=20 x 10 cm	0.15	0.11	0.06	0.04	0.03	0.06
T_6=20 x 5 cm	0.14	0.10	0.04	0.06	0.05	0.06
Mean	0.16	0.13	0.11	0.07	0.06	

	Date of sowing (A)	Seed rate/Plant density (B)	A × B
S.Em.±	0.0025	0.0027	0.0062
CD at 5%	0.0071	0.0078	0.0175
Significance	**	**	**

sowing. The response to 40 × 10 cm population density was best at all the stages followed by 40 × 5 cm as well as 30 × 10 cm, respectively (Tables 1.68–1.70).

As far as the interaction effect was concerned, May 5 sowing with 30 × 10 cm spacing as well as May 5 sowing with 30 × 5 cm spacing interactions proved superior compared to other combinations at all growth stages studied (Tables 1.68–1.70).

1.5.2.2 Leaf Nutrient (NPK) Content

Leaf nutrient (NPK) content was significantly affected due to change in date of sowing, population density, and their interaction at all the three growth stages except interaction effect for leaf nitrogen at all growth stages (Tables 1.71–1.73), effect of population density and leaf potassium percentage at 90 days (Table 1.79), and interaction effect of leaf potassium percentages at all the three growth stages (Tables 1.71–1.79). The data are briefly described in the following subsections.

TABLE 1.71 Effect of Seed Rate and Dates of Sowing on Leaf Nitrogen (%) of *Arachis hypogaea* L. (Groundnut) at 50 Days

Seed rate/ plant density	Date of sowing					
	5 May	5 June	5 July	5 August	5 September	Mean
T_1 = 40 x 10 cm	3.22	3.10	3.05	2.80	2.70	2.91
T_2 = 40 x 5 cm	3.19	3.05	3.02	2.76	2.68	2.88
T_3 = 30 x 10 cm	3.24	3.12	3.00	2.70	2.56	2.85
T_4 = 30 x 5 cm	3.26	3.11	3.01	2.68	2.50	2.83
T_5 = 20 x 10 cm	3.01	3.00	2.90	2.60	2.46	2.74
T_6 = 20 x 5 cm	3.05	2.95	2.88	2.50	2.44	2.69
Mean	3.16	3.06	2.98	2.67	2.56	

	Date of sowing (A)	Seed rate/Plant density (B)	A × B
S.Em.±	0.035	0.038	0.086
CD at 5%	0.100	0.109	0.245
Significance	**	**	NS

TABLE 1.72   **Effect of Seed Rate and Dates of Sowing on Leaf Nitrogen (%) of** *Arachis hypogaea* L. (Groundnut) **at 70 Days**

Seed rate/plant density	Date of sowing					
	5 May	5 June	5 July	5 August	5 September	Mean
T_1=40 x 10 cm	3.18	3.05	2.98	2.70	2.48	2.80
T_2=40 x 5 cm	3.11	2.99	2.86	2.66	2.40	2.73
T_3=30 x 10 cm	3.20	3.08	2.84	2.60	2.40	2.73
T_4=30 x 5 cm	3.22	3.20	2.80	2.58	2.38	2.74
T_5=20 x 10 cm	2.98	2.96	2.76	2.55	2.32	2.65
T_6=20 x 5 cm	2.99	2.90	2.74	2.50	2.30	2.61
Mean	3.11	3.03	2.83	2.60	2.38	

	Date of sowing (A)	Seed rate/Plant density (B)	A × B
S.Em.±	0.028	0.030	0.068
CD at 5%	0.079	0.087	0.194
Significance	**	**	NS

TABLE 1.73   Effect of Seed Rate and Dates of Sowing on Leaf Nitrogen (%) of *Arachis hypogaea* L. (groundnut) at 90 Days

Seed rate/plant density	Date of sowing					
	5 May	5 June	5 July	5 August	5 September	Mean
T1 = 40 x 10 cm	2.90	2.70	2.50	2.30	2.10	2.40
T2 = 40 x 5 cm	2.86	2.68	2.44	2.28	2.11	2.38
T3 = 30 x 10 cm	2.90	2.60	2.40	2.16	2.05	2.30
T4 = 30 x 5 cm	2.92	2.58	2.38	2.10	2.11	2.29
T5 = 20 x 10 cm	2.80	2.50	2.34	2.08	1.98	2.23
T6 = 20 x 5 cm	2.75	2.48	2.32	2.04	1.90	2.19
Mean	2.86	2.59	2.40	2.16	2.04	

	Date of sowing (A)	Seed rate/Plant density (B)	A × B
S.Em.±	0.0312	0.0342	0.0765
CD at 5%	0.0884	0.0968	0.216
Significance	**	**	NS

TABLE 1.74   Effect of Seed Rate and Dates of Sowing on Leaf Phosphorus (%) of *Arachis hypogaea* L. (Groundnut) at 50 Days

Seed rate/plant density	Date of sowing					Mean
	5 May	5 June	5 July	5 August	5 September	
T_1=40 x 10 cm	0.56	0.48	0.46	0.40	0.34	0.42
T_2=40 x 5 cm	0.54	0.46	0.44	0.38	0.30	0.40
T_3=30 x 10 cm	0.55	0.52	0.46	0.38	0.33	0.42
T_4=30 x 5 cm	0.56	0.52	0.48	0.38	0.35	0.43
T_5=20 x 10 cm	0.52	0.44	0.40	0.33	0.30	0.37
T_6=20 x 5 cm	0.50	0.42	0.38	0.30	0.28	0.35
Mean	0.54	0.47	0.44	0.36	0.32	
	Date of sowing (A)		Seed rate/Plant density (B)		A × B	
S.Em.±	0.0073		0.0080		0.0180	
CD at 5%	0.0209		0.0228		0.0512	
Significance	**		**		NS	

TABLE 1.75   Effect of Seed Rate and Dates of Sowing on Leaf Phosphorus (%) of *Arachis hypogaea* L. (Groundnut) at 70 Days

Seed rate/plant density	Date of sowing					Mean
	5 May	5 June	5 July	5 August	5 September	
T_1=40 x 10 cm	0.56	0.48	0.48	0.44	0.38	0.45
T_2=40 x 5 cm	0.56	0.46	0.46	0.42	0.32	0.42
T_3=30 x 10 cm	0.58	0.55	0.48	0.42	0.36	0.45
T_4=30 x 5 cm	0.58	0.53	0.52	0.44	0.36	0.46
T_5=20 x 10 cm	0.56	0.50	0.44	0.36	0.34	0.41
T_6=20 x 5 cm	0.54	0.48	0.42	0.34	0.34	0.40
Mean	0.56	0.50	0.47	0.40	0.35	
	Date of sowing (A)		Seed rate/Plant density (B)		A × B	
S.Em.±	0.0048		0.0052		0.0117	
CD at 5%	0.0136		0.0149		0.0333	
Significance	**		**		**	

TABLE 1.76　Effect of Seed Rate and Dates of Sowing on Leaf Phosphorus (%) of *Arachis hypogaea* L. (Groundnut) at 90 Days

Seed rate/plant density	Date of sowing					Mean
	5 May	5 June	5 July	5 August	5 September	
T1 = 40 x 10 cm	0.60	0.54	0.52	0.48	0.40	0.49
T2 = 40 x 5 cm	0.59	0.52	0.48	0.40	0.36	0.44
T3 = 30 x 10 cm	0.62	0.60	0.52	0.43	0.40	0.49
T4 = 30 x 5 cm	0.61	0.60	0.50	0.46	0.40	0.49
T5 = 20 x 10 cm	0.58	0.54	0.48	0.40	0.38	0.45
T6 = 20 x 5 cm	0.60	0.52	0.44	0.36	0.36	0.42
Mean	0.60	0.55	0.49	0.42	0.38	

	Date of sowing (A)	Seed rate/Plant density (B)	A × B
S.Em.±	0.0045	0.0050	0.0111
CD at 5%	0.0129	0.0141	0.0316
Significance	**	**	**

TABLE 1.77　Effect of Seed Rate and Dates of Sowing on Leaf Potassium (%) of *Arachis hypogaea* L. (Groundnut) at 50 Days

Seed rate/plant density	Date of sowing					Mean
	5 May	5 June	5 July	5 August	5 September	
T_1 = 40 x 10 cm	1.68	1.60	1.42	1.30	1.28	1.40
T_2 = 40 x 5 cm	1.64	1.58	1.40	1.28	1.22	1.37
T_3 = 30 x 10 cm	1.80	1.69	1.48	1.30	1.20	1.42
T_4 = 30 x 5 cm	1.82	1.69	1.52	1.34	1.22	1.44
T_5 = 20 x 10 cm	1.62	1.58	1.40	1.25	1.20	1.36
T_6 = 20 x 5 cm	1.60	1.56	1.38	1.24	1.18	1.34
Mean	1.69	1.62	1.43	1.29	1.22	

	Date of sowing (A)	Seed rate/Plant density (B)	A × B
S.Em.±	0.025	0.028	0.063
CD at 5%	0.073	0.080	0.179
Significance	**	*	NS

TABLE 1.78   Effect of Seed Rate and Dates of Sowing on Leaf Potassium (%) of *Arachis hypogaea* L. (Groundnut) at 70 Days

Seed rate/plant density	Date of sowing					Mean
	5 May	5 June	5 July	5 August	5 September	
T_1 = 40 x 10 cm	1.82	1.70	1.50	1.44	1.30	1.49
T_2 = 40 x 5 cm	1.84	1.68	1.52	1.38	1.28	1.47
T_3 = 30 x 10 cm	1.89	1.76	1.52	1.36	1.26	1.48
T_4 = 30 x 5 cm	1.88	1.78	1.50	1.34	1.24	1.47
T_5 = 20 x 10 cm	1.76	1.68	1.44	1.32	1.22	1.42
T_6 = 20 x 5 cm	1.74	1.65	1.42	1.30	1.24	1.40
Mean	1.82	1.71	1.48	1.36	1.26	

	Date of sowing (A)	Seed rate/Plant density (B)	A × B
S.Em.±	0.027	0.029	0.066
CD at 5%	0.077	0.084	0.188
Significance	**	NS	NS

TABLE 1.79   Effect of Seed Rate and Dates of Sowing on Leaf Potassium (%) of *Arachis hypogaea* L. (Groundnut) at 90 Days

Seed rate/plant density	Date of sowing					Mean
	5 May	5 June	5 July	5 August	5 September	
T_1 = 40 x 10 cm	1.82	1.74	1.52	1.48	1.40	1.54
T_2 = 40 x 5 cm	1.86	1.70	1.52	1.44	1.30	1.49
T_3 = 30 x 10 cm	1.92	1.79	1.50	1.42	1.28	1.50
T_4 = 30 x 5 cm	1.90	1.80	1.54	1.38	1.28	1.50
T_5 = 20 x 10 cm	1.80	1.68	1.44	1.34	1.26	1.43
T_6 = 20 x 5 cm	1.78	1.66	1.44	1.32	1.22	1.41
Mean	1.82	1.74	1.52	1.48	1.40	1.54

	Date of sowing (A)	Seed rate/Plant density (B)	A × B
S.Em.±	0.033	0.036	0.081
CD at 5%	0.094	0.103	0.231
Significance	**	NS	NS

1.5.2.2.1   Leaf Nitrogen (%)

Leaf nitrogen percentage was significantly affected due to change in date of sowing, row to plant distances, and their interaction effect at all growth stages (Tables 1.71–1.73).

The effect of May 5 sowing was best; the lowest value was recorded in September 5 at all growth stages. The response of population density was highest in 40 × 10 cm sowing, followed by 40 × 5 cm as well as 30 × 10 cm population density, respectively at all growth stages (Tables 1.71–1.73).

1.5.2.2.2   Leaf Phosphorus (%)

Leaf phosphorus percentage was significantly affected due to change in date of sowing, population density, and their interaction at all growth stages (Tables 1.74–1.76). The effect of May 5 sowing was best; the least was recorded in September 5 sowing at all growing stages. The response of 30 × 5 cm population density gave maximum value, statistically equal to 30 × 10 cm followed by 40 × 10 cm population density at all growth stages.

Among the interaction, May 5 sowing with 30 × 5 cm spacing as well as May 5 sowing with 30 × 10 cm spacing proved superior and best compared to the rest of the interactions (Tables 1.74–1.75).

1.5.2.2.3   Leaf Potassium (%)

Leaf potassium percentage was significantly affected, except for the response to population density at 90 days and interaction effect at all growth stages (Tables 1.77–1.79), due to change in the date of sowing, population density, and their interaction at all growth stages. The effect of May 5 sowing was best; it was lowest in September 5 at all growth stages.

The response to 30 × 5 cm population density was best, followed by 30 × 10 cm as well as 40 × 10 cm population density, respectively both at 50 and 70 days (Tables 1.77–1.79).

1.5.2.3 Yield Characteristics

At harvest, except the response to population density and interaction effect of pod weight (Table 1.83) and seed oil (%), yield characteristics (plant weight, number and weight of pod, total pod yield, and oil content (%) of seed) were significantly affected due to change in the date of sowing, row to plant distances, and their interaction (Tables 1.80–1.86). The results are briefly described separately in the following subsections.

1.5.2.3.1 Weight/Plant (g)

At harvest, plant weight was significantly affected due to change in date of sowing, row to plant spacing, and their interactions (Table 1.81). The effect was highest in May 5 sowing and lowest in September 5 sowing. The response to 40 × 10 cm population density was maximum, followed by 40 × 5 cm and 30 × 10 cm population density, respectively.

As far as the interaction effect was concerned, May 5 sowing with 40 × 10 cm spacing was best, statistically equal to May 5 sowing with 30 × 10 cm population density (Table 1.81).

TABLE 1.80 Effect of Seed Rate and Dates of Sowing on Fresh Weight (g/plant) of *Arachis hypogaea* L. (Groundnut) at Harvest

Seed rate/plant density	Date of sowing					Mean
	5 May	5 June	5 July	5 August	5 September	
T$_1$ = 40 x 10 cm	362.42	303.67	258.00	245.00	221.00	256.92
T$_2$ = 40 x 5 cm	336.76	294.00	251.00	239.67	199.00	245.92
T$_3$ = 30 x 10 cm	361.75	278.34	250.34	233.34	203.66	241.42
T$_4$ = 30 x 5 cm	298.43	266.34	242.67	225.32	202.35	234.17
T$_5$ = 20 x 10 cm	248.42	268.33	233.33	219.00	194.00	228.67
T$_6$ = 20 x 5 cm	240.00	220.00	196.65	169.35	163.68	187.42
Mean	307.96	271.78	238.67	221.95	197.28	

	Date of sowing (A)	Seed rate/Plant density (B)	A × B
S.Em.±	4.45	4.87	10.90
CD at 5%	12.60	13.80	30.86
Significance	**	**	**

TABLE 1.81 Effect of Seed Rate and Dates of Sowing on Pod Number/Plant of *Arachis hypogaea* L. (Groundnut) at Harvest

Seed rate/plant density	Date of sowing					Mean
	5 May	5 June	5 July	5 August	5 September	
T_1 = 40 x 10 cm	33.01	25.34	27.07	21.09	19.01	23.13
T_2 = 40 x 5 cm	33.02	28.67	26.14	20.08	11.12	21.50
T_3 = 30 x 10 cm	30.32	20.32	23.15	17.21	10.08	17.69
T_4 = 30 x 5 cm	30.05	19.36	23.17	09.23	08.07	21.27
T_5 = 20 x 10 cm	27.16	17.65	18.30	09.16	08.01	17.98
T_6 = 20 x 5 cm	23.25	12.00	13.23	07.05	07.06	12.62
Mean	29.47	20.56	21.84	19.46	13.40	

	Date of sowing (A)	Seed rate/Plant density (B)	A × B
S.Em.±	0.316	0.346	0.774
CD at 5%	0.895	0.980	2.192
Significance	**	**	**

TABLE 1.82 Effect of Seed Rate and Dates of Sowing on Weight of Pods (g/plant) of *Arachis hypogaea* L. (Groundnut) at Harvest

Seed rate/plant density	Date of sowing					Mean
	5 May	5 June	5 July	5 August	5 September	
T_1 = 40 x 10 cm	45.34	34.27	23.24	14.20	9.32	20.26
T_2 = 40 x 5 cm	45.22	34.57	23.80	12.57	9.41	20.09
T_3 = 30 x 10 cm	44.64	33.53	24.47	13.60	9.30	20.23
T_4 = 30 x 5 cm	44.86	33.87	22.80	12.04	8.62	19.33
T_5 = 20 x 10 cm	43.43	33.23	23.50	10.00	7.44	18.54
T_6 = 20 x 5 cm	43.42	32.73	22.57	11.57	6.43	18.33
Mean	44.49	33.70	23.40	12.33	8.42	

	Date of sowing (A)	Seed rate/Plant density (B)	A × B
S.Em.±	0.67	0.74	1.66
CD at 5%	1.92	2.10	4.70
Significance	**	NS	NS

1.5.2.3.2   Number of Pod/Plant

Pod number was significantly affected due to change in date of sowing, population density, and their interaction (Table 1.82). The effect of May 5 sowing was best on this yield attribute and lowest in September 5 sowing. The response to 40 × 10 cm population density was highest, statistically equal to 30 × 5 cm followed by 40 × 5 cm population density.

As far as the interaction effect was concerned, May 5 sowing with 30 × 10 cm spacing as well as May 5 sowing with 30 × 5 cm spacing interaction proved superior when compared to the rest of the combinations (Table 1.81).

1.5.2.3.3   Weight of Pod/Plant (g)

Weight of pod was significantly affected due to change in the date of sowing (Table 1.82). Maximum value was recorded in May 5 sowing and minimum was noted in September 5 sowing.

1.5.2.4   Pod Yield (kg/ha)

At harvest, total pod yield was significantly affected due to change in date of sowing, population density, and their interaction (Table 1.83). The effect on total pod yield was maximum in May 5 sowing, statistically equal to June 5 sowing; minimum pod yield was recorded in September 5 sowing. It was clearly noted that the response to 30 × 10 cm population density was best, statistically equal to 40 × 5 cm followed by 30 × 5 cm spacing, respectively.

In addition, 9.7% more total pod yield was found in 30 × 10 cm spacing as compared to 30 × 5 cm population density (Table 1.83).

Regarding interaction effect, May 5 sowing with 30 × 5 cm spacing, May 5 sowing with 30 × 10 cm spacing, and June 5 combination proved superior and statistically equal to each other for this important parameter (Table 1.83).

1.5.2.5   Total Oil Content (%)

The total seed oil percentage was significantly affected due to change in population density (row to plant distances) parameter (Table 1.84).

TABLE 1.83 Effect of Seed Rate and Dates of Sowing on Pod Yield (kg/ha) of *Arachis hypogaea* L. (Groundnut) at Harvest

Seed rate/plant density	Date of sowing					Mean
	5 May	5 June	5 July	5 August	5 September	
T_1 = 40 x 10 cm	1925.00	1840.00	1630.00	1500.00	1215.00	1546.25
T_2 = 40 x 5 cm	1860.00	1840.00	1745.00	1620.00	1245.00	1612.50
T_3 = 30 x 10 cm	1990.00	1875.00	1885.00	1670.00	126.00	1389.00
T_4 = 30 x 5 cm	1998.00	1875.00	1545.00	1620.00	1185.00	1556.25
T_5 = 20 x 10 cm	1745.00	1645.00	1580.00	1440.00	1010.00	1418.75
T_6 = 20 x 5 cm	1500.00	1385.00	1305.00	1120.00	975.00	1196.25
Mean	1836.33	1743.33	1615.00	1495.00	959.33	

	Date of sowing (A)	Seed rate/Plant density (B)	A × B
S.Em.±	34.54	37.84	84.61
CD at 5%	97.78	107.12	239.53
Significance	**	**	**

Maximum seed oil percentage was noted in 40 × 10 cm population density, statistically equal to 40 × 5 cm and 30 × 10 cm spacing, respectively (Table 1.84).

1.6 DATA INTERPRETATION AND JUSTIFICATION IN CONTEXT TO GROUNDNUT AGRICULTURE

1.6.1 GENERAL

The profitability of the groundnut crop can be increased through better management practices, including timely sowing, proper plant density, balanced fertilization, optimum scheduling of irrigation and also require appropriate plant protection measures. On groundnut (*Arachis hypogaea* L.), a perusal of the following publication: Fortanier (1957); Holley and Hammons (1968); Wood (1968); Reddy et al. (1981); Ball et al. (1983); Ketring (1984); Singh and Ahuja (1984); Ramesh Babu et al. (1985); Panchaksharaiah (1985); Selamat and Gardner (1985); Ravi Kumar et al. (1994); Chawale et al. (1995); Patra et al. (1995); Varalakshmi et al.

TABLE 1.84   Effect of Seed Rate and Dates of Sowing on Total Seed Oil (%) of *Arachis hypogaea* L. (Groundnut) at Harvest

Seed rate/plant density	Date of sowing					Mean
	5 May	5 June	5 July	5 August	5 September	
T1 = 40 x 10 cm	51.71	50.30	50.46	49.74	49.12	49.91
T2 = 40 x 5 cm	49.70	49.31	48.72	48.42	48.23	48.67
T3 = 30 x 10 cm	48.45	48.00	48.45	49.06	47.32	48.21
T4 = 30 x 5 cm	48.72	45.72	46.00	46.05	47.25	46.26
T5 = 20 x 10 cm	47.71	47.00	48.43	48.08	49.01	48.13
T6 = 20 x 5 cm	47.33	47.00	48.46	48.07	47.07	47.65
Mean	48.94	47.89	48.42	48.24	48.00	

	Date of sowing (A)	Seed rate/Plant density (B)	A × B
S.Em.±	0.624	0.684	1.529
CD at 5%	1.788	1.936	4.331
Significance	NS	*	NS

(2005); Yakabri and Satyanarayan (1995); Sarkar et al. (1998); Talwar et al. (1999); Nalawade and Patil (2000); Murthy et al. (2000); Kachot et al. (2001); Kumar et al. (2003); Hadwani and Gundalia (2005); Ahmad et al. (2007); Carley et al. (2008); Mohapatra and Dixit (2010) and other publications highlights the importance of the present study.

It may not be out of place to mention here that the work done at Shahjahanpur, during the last two decades or so, on the same important aspects of applied physiology of cereals, oil crops, medicinal, and sugar crops under the supervision of Dr. Abbas and others has yielded valuable results [Abbas et al. (1980); Abbas and Kumar (1987); Kumar and Abbas (1992); Kishor and Abbas (2003); Kishor (2006); Kishor et al. (2006); Hasan and Abbas (2007a,b) and (2008); Hasan et al. (2007, 2009); Kumar et al. (2007, 2008a,b); Kumar et al. (2008a,b); Sharma (2007); Kanaujia (2008); Kumar (2008); Verma (2011); Kumar et al. (2010) Shukla (2010); Sajjad et al. (2011); and Kumar et al. (2011); Kumar (2012)].

The inclusion of an additional bonus crop with improved agro-techniques under the local cropping pattern has been carried further using groundnut (*Arachis hypogaea* L.) by the present author. Thus, the scope of groundnut cultivation has been enlarged as is brought out by the following brief discussion of the results obtained from the two field trials undertaken during 2008 to 2010, laying particular emphasis on yield.

1.6.2   RESULTS OF EXPERIMENT 1

Economics of production would ultimately decide the profitability of groundnut oil production, and optimization of inputs would be a necessary pre-requisite if groundnut production has to be adopted with economic viability in the country and local area, particularly as a bonus crop. Application of adequate quantities of manures and fertilizers is considered to be one of the most important factors for increasing groundnut yield. In order to get an idea of the rate at which various nutrients should be applied, it is essential to be familiar with the response of the crop to the method of application and source of the nutrient under different soil conditions.

Let us summarize our findings from the first field trial. It was observed that the highest pod yield was noted in 3/4 soil + 1/4 foliar application of nitrogen, followed by 2/3 soil + 1/3 foliar application (Table 1.44), and lowest yield was recorded in full foliar application of nitrogen (Table 1.44). As far as the seed oil content (%) in groundnut was concerned, the highest percentage was found in full soil application, statistically equal to 3/4 soil + 1/4 foliar application as well as 2/3 soil + 1/3 foliar application, while lowest oil percentage was given by full foliar application of nitrogen (Table 1.45). This increase in pod yield and oil percentage was associated with increased leaf number (Table 1.11–1.13), fresh weight of plant (Table 1.41), pod number (Table 1.42), and pod weight (Table 1.43) as well as due to increased leaf nitrogen (Table 1.32–1.34) and leaf potassium (Tables 1.38–1.40) percentages. Thus, increase in the yield with nitrogen application could be ascribed to the overall improvement in plant growth and vigor as it plays an important role in plant metabolism, resulting in better yield attributes and yields. The results are in accordance with Patel et al. (1994) and Hadwani and Gundalia (2005). The highest pod yield (8.5% and more) was noted in 3/4 soil + 1/4 foliar nitrogen application as compared to full soil nitrogen application (Table 1.44). Similarly, the seed oil percentage was also statistically equal in full soil application as well as

3/4 soil + 1/4 foliar application (Table 1.45). The superiority of foliar nutrition might be due to coincidence of foliar application, as a supplement to soil application, with peak nutrition requirement of the crop. The quantity of nutrients absorbed by the roots at this peak period of nutrient requirement may not be sufficient to meet the needs at pod development stage, and supplementing nutrients through foliage might have resulted in better nutrient balance in the plants leading to increased yield components. Similar results were also observed by Reddy et al. (1991), Patra et al. (1995), and Panchakshariah (1985). The leaf number (Tables 1.11–1.13) was also high in these high-yielding treatments. Nitrogen application increases crop growth rate, and leaf area index significantly as has also been observed by Selamat and Gardner (1985). Thus, the response of growth to nitrogen might be attributed to rapid meristematic activity due to increased leaf nutrient percentages (Tables 1.32–1.40) at all stages of growth and development in groundnut (*Arachis hypogaea* L.) as well as due to formation of healthy pegs (developing fruit) that absorb more nutrients from the fruiting zone for better filling up of pods.

As far as the interactions were concerned, 2/3 soil + 1/3 foliar application of nitrogen in the form of ammonium sulfate was best for pod yield (Table 1.44), statistically equal to 1/3 soil + 2/3 foliar application of nitrogen in the form of diammonium phosphate. The seed oil percentage was also maximum in 2/3 soil + 1/3 foliar application of nitrogen in the form of ammonium sulfate (Table 1.45). This clearly showed the superiority of spraying of plant nutrients after soil application, leading to enhanced pod yield and oil percentage in groundnut associated with higher growth (Tables 1.11–1.13), leaf nutrient (NPK) content (Tables 1.32–1.40), and yield parameters (Tables 1.41–1.45). An almost similar trend was also reported by Patra et al. (1995) in groundnut.

1.6.3 RESULTS OF EXPERIMENT 2

Let us summarize our findings from the second field trial. It was found that May 5 sowing significantly produced highest pod yield, statistically equal to June 5 sowing (Table 1.83). The lowest pod yield was noted in September 5 sowing. May 5 sowing was associated with maximum values in all the parameters of growth, nutrient (NPK) content, photosynthetic pigments, yield attributes (Tables 1.47–1.82), and their cumulative effect; it could be ascribed to the effectiveness of the crop to exploit this favorable environment condition toward yield formation. Due to the groundnut's indeterminate growth characteristics, it has been widely assumed that at later stages, both vegetative and reproductive structures compete with each other for nutrients and metabolites. Thus, higher availability of these to sink might have reduced the competition between developing structures, which consequently resulted in improved pods/plant (Table 1.81) under favorable suitable planting. These findings are in agreement with those reported by Gill and Kumar (1995). Groundnut (*Arachis hypogaea* L.) can be grown over a wide range of climatic conditions and the advantages of favorable weather conditions, particularly temperature in this crop are well known (Kumar, 2012; Kumar et al., 2003; Carley et al., 2008; Talwar et al., 1999). The population density variation also significantly affected pod yield as well as seed oil percentage (Tables 1.83 and 1.84). Highest pod yield was noted in 30 × 10 cm sowing, statistically equal to 40 × 5 cm sowing. The seed oil percentage was maximum in 40 × 5 cm population density, statistically equal to 40 × 5 cm, as well as 30 × 5 cm. These results show the extreme plasticity of the crop due to the intense competition faced by the individuals and the acute competitive ability of the crop. The results corroborate the findings of Murthy and Rao (2000), and Singh and Ahuja (1985).

During interaction, a maximum pod yield was noted in 30 × 5 cm spacing with June 5 sowing, statistically equal to 30 × 5 cm spacing with May 5 sowing (Table 1.83). This was associated with increased plant height (Tables 1.53–1.55), dry weight (Tables 1.59–1.61), photosynthetic pigments (Tables 1.62–1.70), leaf phosphorus content (Tables 1.74–1.76) due to the plant exploiting the most favorable environment leading to increased photosynthesis, translocation, and membrane permeability (Björkman et al., 1980).

1.6.4   PROPOSALS FOR FUTURE WORK

It is evident from this discussion that some of the problems related to groundnut crop husbandry regarding source and method of nitrogen application; impact of population density and date of sowing to improve agro-techniques for inclusion in the local cropping pattern can be solved, and an additional bonus crop can be generated between the rabi and kharif season under local conditions. This will help the farmers to generate more income and profit leading to their better welfare.

The effect of various organic manures, biofertilizers along with new emerging methods (sprinkler system) of irrigation can be tested alone or in multiple cropping systems.

Further, mere increase in yield would be meaningless unless the content of active principles and oil quality is also tested and improved under different agro-climatic conditions/stresses.

KEYWORDS

- *Arachis hypogaea* L.
- date of sowing
- groundnut
- nitrogen
- seed rate

CHAPTER 2

INTEGRATED NUTRIENT MANAGEMENT

CONTENTS

2.1   INTRODUCTION

Groundnut (*Arachis hypogaea* L.), also known as peanut especially in the USA, is a herbaceous annual plant of the family Leguminosae (Fabaceae). It is grown for its oil and edible nuts and is a crop of worldwide significance. Plants are small, usually erect, and thin stemmed with feather-like leaves grown by both marginal and large commercial producers. The leaves are arranged in alternate pairs. It produces yellow, orange, cream, or white flowers, which produces "pegs" (developing fruit). Pegs mature into the ground to grow the pod. The pods can reach up to 10 cm in length and can contain 1–5 seeds. The plant can reach 0.6 m (2 ft) in height depending on the variety and lives for only one growing season. It is grouped as both a grain legume and an oil crop due to its high oil content. The scientific classification is as follows:

Kingdom	:	Plantae
Division	:	Angiosperms
Class	:	Dicotyledons
Sub-class	:	Polypetalae
Series	:	Calyciflorae
Order	:	Rosales
Family	:	Leguminosae
Sub-family	:	Papilionatae
Genus	:	Arachis
Species	:	hypogaea
Bionomial name	:	*Arachis hypogaea* L.

This crop grows best in light sandy loam soil with a pH of 6.0–7.0 and very well in tropical and subtropical climates. It requires warm temperatures and a long growing season. Plants grow optimally at high humidity and a temperatures between 30 and 34°C. Temperatures above 34°C may damage flowers. They can grow with as little as 350 mm of water but for optimum production, a rainfall between 500 and 600 mm water over the course of the growing season is ideal. Groundnut plants continue to produce flowers when pods are developing, therefore, even when they are ready for harvest, some pods are immature. The timing of harvest is an important decision to maximize yield. If it is too early, many pods will be unripe. If too late, the pods will become loose at the stalk and remain in the soil (Anonymous, 2008). The plants are pulled and inverted manually or by machines leaving the plant upside down on the ground. They are left for 6 to 8 days exposed to the sun to remove about 1/3 of their original moisture level. The pods are threshed, dried properly, and stored in dry conditions otherwise they may become infected by the mold fungus *Aspergillus flavus,* which releases a toxic and highly carcinogenic substance, aflatoxin.

The bacterial, fungal, nematodes, phytoplasma, viruses or pests diseases that commonly infect groundnut are given as follows:

Bacterial diseases
Bacterial wilt   *Pseudomonas solanacearum*
Fungal diseases
Alternaria leaf blight   *Alternaria tenuissima*

Alternaria leaf spot	*Alternaria arachidis*
Alternaria spot and veinal necrosis	*Alternaria alternata*
Anthracnose	*Colletorichum arachidis*
	Colletorichum dematium
	Colletotrichum mangenoti
Aspergillus crown rot	*Aspergillus niger*
Blackhull	*Thielaviopsis basicola*
	Chalara elegans [synanmorph]
Botrytis blight	*Botrytis cinerea*
	Botryotinia fuckeliana [telemorph]
Charcoal rot and Macrophomina leaf spot	*Macrophomina phaseoiina*
	=Rhizoctonia baticola
Choanephora leaf spot	*Choanephora spp.*
Collar rot	*Lasiodiplodia theobromae*
	Diplodia gossypina
Colletotrichum leaf spot	*Colletotrichum gloeosporioides*
	Glomerella cingulata [teleomorph]
Cylindrocladium black rot	*Cylindrocladium crotalariae*
	Calonectria crotalariae [telemorph]
Cylindrocladium leaf spot	*Cylindrocladium scoparium*
	Calonectria kyotensis [telemorph]
Damping-off, Aspergillus	*Aspergillus flavus*
	Aspergillus niger
Damping-off, Fusarium	*Fusarium spp.*
Damping-off, Pythium	*Pythium spp.*
Damping-off, Rhizoctonia	*Rhizoctonia spp.*
Damping-off, Rhizopus	*Rhizopus spp.*
Drechslera leaf spot	*Bipolaris spicifera*
	Drechslera spicifera
	Cochliobolus spicifer [teleomorph]
Fusarium peg and root rot	*Fusarium spp.*
Fusarium wilt	*Fusarium oxysporum*
Leaf spot, early	*Cercospora arachidicola*
	Mycosphaerella arachidis [teleomorph]
Leaf spot, late	*Phaeosariopsis personata*

	Cercosporidium personatum
	Mycosphaerella berkeleyi
	[teleomorph]
Melanosis	*Stemphylium botryosum*
	Pleospora tarda [teleomorph]
Myrothecium leaf blight	*Myrothecium roridum*
Olpidium root rot	*Olpidium brassicae*
Pepper spot and scorch	*Leptosphaerulina crassiasca*
Pestalotiopsis leaf spot	*Pestalotiopsis arachidis*
Phoma leaf blight	*Phoma microspora*
Phomopsis foliar blight	*Phomopsis phaseoli*
	Phomopsis sojae
	Diaporthe phaseolorum
	[teleomorph]
Phomopsis leaf spot	*Phomopsis spp.*
Phyllosticta leaf spot	*Phyllosticta arachidis-hypogaeae*
	Phyllosticta sojaecola
	Pleosphaerulina sojicola
	[teleomorph]
Phymatotrichum root rot	*Phymatotrichopsis omnivora*
	Phymatotrichum omnivorum
Pod rot (pod breakdown)	*Fusarium equiseti*
	Fusarium scirpi
	Gibberella intricans [teleomorph]
	Fusarium solani
	Nectria haematococca [telemorph]
	Pythium myriotylum
	Rhizoctonia solani
	Thanatephorus cucumeris
	[teleomorph]
Powdery mildew	*Oidium arachidis*
Pythium peg and root rot	*Pythium myriotylum*
	Pythium aphanidermatum
	Pythium debaryanum
	Pythium irregulare
	Pythium ultimum

Pythium wilt	*Pythium myriotylum*
Rhiozoctonia foliar blight, peg and root rot	*Rhizoctonia solani*
Rust	*Puccinia arachidis*
Scab	*Sphaceloma arachidis*
Sclerotinia blight	*Sclerotinia minor*
	Sclerotinia sclerotiorum
Stem rot (southern blight)	*Sclerotium rolfsii*
	Athelia rolfsii [teleomorph]
Verticillium wilt	*Verticillium albo-atrum*
	Verticillium dahlia
	Phoma arachidicola
	Ascochyta adzamethica
Web blotch (net blotch)	*Didymosphaeria arachidicola*
	Mycosphaerella arachidicola
Yellow mold	*Aspergillus flavus*
	Aspergillus parasiticus
Zonate leaf spot	*Cristulariella moricola*
	Sclerotium cinnamomi [synamorph]
Zonate leaf spot	*Grovesinia pyramidalis* [teleomorph]

Nematodes, parasitic

Dagger	*Xiphinema spp.*
Pod lesion	*Tylenchorhynchus brevilineatus*
	Tylenchorhynchus brevicadatus
Ring	*Criconemella ornata*
	Macroposthonia ornata
Root-knot, Javanese	*Meloidogyne javanica*
Root-knot, northern	*Meloidogyne hapla*
Root-knot, peanut	*Meloidogyne arenaria*
Root lesion	*Pratylenchus brachyurus*
	Pratylenchus coffeae
Seed and pod	*Ditylenchus destructor*
Spiral	*Scutellonema cavenessi*
Sting	*Belonolaimus glacilis*
	Belonolaimus longicaudatus
Testa	*Aphelenchoides arachidis*

Phytoplasma, Virus and virus like diseases

Cowpea mild mottle	*Cowpea mild mottle virus*
Groundnut crinkle	*Groundnut crinkle virus*
Groundnut eyespot	*Groundnut eyespot virus*
Groundnut rosette	*Groundnut chlorotic rosette virus*
	Groundnut green rosette virus
Groundnut streak	*Groundnut streak virus*
Marginal chlorosis	Unknown (virus like)
Peanut clump	*Peanut clump virus*
Peanut green mosaic	*Peanut green mosaic virus*
Peanut mottle	*Peanut mottle virus*
Peanut ringspot or bud necrosis	*Tomato spotted wilt virus*
Peanut strip	*Peanut strip virus*
Peanut stunt	*Peanut stunt virus*
Peanut yellow mottle	*Peanut yellow mottle virus*
Tomato spotted wilt	*Tomato spotted wilt virus*
Witches broom	*Phytoplasma*

Pest diseases

Army worms	*Spodoptera frugiperda*
Gram pod borer	*Helicoverpa armigra*
Groundnut aphid	*Aphis craccivora*
Groundnut white grub	*Holotrichia consanguinea*
Jassids	*Empoasca kerri bachlucha* spp.
Leaf miner	*Aproaerema modicella*
Leaf Webber	*Anarsia ephippias* Termite
	Odontotermes obasus
Thrips	*Scirtothrips dorsalis*
	Thrips palmi
Tobacco caterpillar	*Spodoptera litura*
Two-spotted spider mite	*Tetranychus urticae*
Velvet bean caterpillar	*Anticarsia gemmatalis*

The groundnut commonly used for crops is an amphidiploid or allo-tetraploid. It has two sets of chromosomes from two different species thought to be *A. duranensis* and *A. ipaensis* (Seijo et al., 2007; Kochert et al., 1996; Moretzsohn et al., 2013; Husted, 1936; Halward et al., 1992).

It may have originated in northwestern Argentina, or in southeastern Bolivia (Karpovickas et al., 1994; 2007). The groundnut was then taken worldwide by European traders. Cultivation is now very widespread in tropical and subtropical regions. In Asia, it has become a common crop and this region is now the largest producer in the world (Anonymous, 2015a).

The major groundnut producing countries are India, China, the U.S., and West Africa. Groundnut is also cultivated in Burma, the East Indies, Nigeria, and in Southern Europe. Of the total global production, China accounts for 37%, Africa for 25%, India for 21%, the Americas for 8%, and Oceania for 6%. Major exporters are India, which accounts for 37% of world exports. Major importers are the Netherlands, which accounts for 17% of world imports, Indonesia for 10%, Mexico for 7%, Germany for 6%, and Russia for 5% (Anonymous, 2015a).

Groundnut is similar in taste and nutritional value to nuts such as walnuts and almonds and eaten in almost identical ways in western cuisines. Raw kernels (seeds) are commonly squashed, roasted, and eaten as a snack food. It is commercially grown for the extraction of the oil, which is used in cooking. The oil content of groundnut possesses properties similar to olive oil. The by-product of oil extraction is a pressed cake that is used as an animal feed and also in the production of its flour.

Vegetable oils consumption in India is continuously rising and has sharply increased in the couple of years touching around 12.4 kg/head/year (Kumar, 2012). This is still lower than the world average consumption of 17.8 kg. The developed western world has a per capita consumption of 44 to 48 kg/year.

According to Pacharne et al. (2016), groundnut is the premier oilseed crop of India, occupying an area of 6.7 million ha and contributing 7.3 million tons toward oilseed production, which is nearly 40% of the total oilseed production. However, this is lagging behind requirement considerably. There is an urgent need to increase its acreage to fill and improve crop yields (Verma, 2011).

The productivity of groundnut (*Arachis hypogaea* L.), a kharif crop, is low primarily due to its cultivation in poor soils with low secondary and micronutrients, besides an inadequate organic matter (Bandyopadhyay et al., 2000; Nambiar and Ghosh, 1984).

Being a leguminous biologically symbiotic nitrogen-fixing crop, it harbors nitrogen-fixing bacteria (*Rhizobia*) in their root nodules and requires less nitrogen than other crops. It is thus incorporated in cropping systems (crop rotation) to improve soil fertility and soil complex ecosystem. However, groundnut is also an exhaustive crop and removes large amount of macro and micronutrients from soil.

None of the sources of nutrients alone can meet the total plant nutrient needs of the crop adequately. Hence, an integrated use of nutrients from chemical, organic manures, and composts is the most efficient way to supply plant nutrients for sustained crop productivity and improved soil fertility (Singh and Singh, 2002).

2.2 BALANCED/ORGANIC NUTRITION SUPPLEMENTS IN PEANUT CROP AGRO-PHYSIOLOGY

Imbalanced and higher doses of inorganic fertilizers, regular pesticides, and herbicides degrade the soil and threaten the stable traditional ecosystem of the soil (Palaniappan and Annadurai, 1999). There is a need to encourage more healthy, economically sound, and eco-friendly farming system (Bhattacharya and Gehlot, 2003). The use of organic nutrients is now viewed as a profitable farming practice with significant advantages for sustainable crop yields. The uses of manures, wastes, biofertilizers, and composts are time-tested production inputs for improving the sustainable productivity of soil. Composting of organic wastes like crop residues, animal wastes, household and industrial wastes, etc. is a traditional practice being adopted by our farmers since ancient times. Organic farming integrated with chemical fertilizers is gaining gradual momentum across the world. Moreover, growing knowledge of health and environmental issues in the agricultural sector has led to the demand for organic food, which is emerging as an attractive source of rural income generation. While consumer demand for organics are becoming discernible, sustainability in cultivation of crops has become the main concern in agricultural development.

Historically, organic manuring was practiced by ancient civilizations like those in Mesopotamia, Hwang Ho basin, etc. —Bhattacharya and

Chakraborty (2005) report that it began about 1000 years old, dating back to the Neolithic age. In the Ramayana, it is noted that all dead things—rotting corpse or sinking garbage—returned to the earth are transformed into wholesome things that nourish life; such is the alchemy of mother earth. In the Mahabharata (5500 BC), there is mention of Kamadhena, the celestial cow and its role on human life and soil fertility. The Kautilya Arthashastra (300 BC) refer to several organic products like oil cake, excreta of animals. In the Rig Veda (2500–1500 BC), it is stated that to cause healthy growth, the plant should be nourished by dung of goat, sheep, water as well as meat. Even, in the Holy Quran (590 AD), it is written that at least one-third of what is taken out from soils must be returned to it, implying recycling or use of post-harvest residue.

Farmyard manure (FYM) is one of the potential sources of plant nutrients in India as a result of the high cattle population—an average of 6.1 cattle per family (Verma, 2011). Currently, there exists a long list of literature reporting the findings of research that reveal the importance (efficiency and effectiveness) of FYM and other organic nutrient sources in maintaining soil fertility, improving crop yields, sustaining productivity, and that display their increased potential when used in integration with fertilizers (Verma, 2011).

Similarly, sugar mill based by-products like pressmud is produced in bulk quantity (about 5 million tons per year) in India (Jambhekar, 1992). Uttar Pradesh itself produces about 1.5 million tons of pressmud every year. Its disposal is a costly problem and if stored, it pollutes the atmosphere in the vicinity (James and Hasibuan, 2002). But it is one of the best sources of organic matter to replenish the soil (Singh et al., 2006). In addition to direct use of pressmud along with gamma-BHC, there is considerable scope in effectively using these wastes by a tightly controlled but low technology composting process to ensure the conservation of the environment for sustainable development (Singh et al., 2006).

This pressmud (compost) yielded good quality compost in a shorter period (75 days), having about 50% more nutritional value and about 8 times more microbial value in comparison to conventional compost (Singh et al., 2006).

Groundnut production too can be increased considerably under local conditions with balanced and integrated use of fertilizer and manure.

Since there is a need to encourage a more productive, cost efficient and eco-friendly farming system (Bhattacharya and Gehlot, 2003) and also keeping in view the relevance of the topic, the present author decided to test *Arachis hypogaea* L. (groundnut) Kausal G-201 variety under local conditions (a Tarai region of western Uttar Pradesh). Two field experiments were conducted in 2008–2009 and 2009–2010 with the following objectives.

1. To study the effect of soil applied pressmud raw with gamma-BHC, pressmud compost and farmyard manure @ 0, 5, 10, 15, and 20 ton/ha on growth, yield, and oil content of groundnut.

2. To study the effect of soil applied split doses of recommended dose of various basally applied NPK (nitrogen, phosphorus and potassium; full and half) along with the best organic source, split and applied at different timings, to get optimum yield as well as in order to save and economize the costly input (fertilizer) if any, on growth, yield, and oil content of the groundnut.

Both the experiments were properly replicated and the data analyzed statistically. The conclusions drawn have been discussed in the light of the findings of the other researchers in this book.

2.3   PRESENT SCENARIO OF THE RESEARCH

2.3.1   REVIEW OF LITERATURE

2.3.1.1   General

As has been mentioned before, groundnut is an exhaustive crop and removes large amount of macro and micronutrients from soil. An integrated use of nutrients from chemical, organic and agricultural wastes is the most efficient way to supply plant nutrients for sustained crop productivity and to improve soil fertility (Singh and Singh, 2002). With this brief background, we will review literature to present the most important and relevant research work done on growth, yield, leaf nutrient content (NPK), and oil content in groundnut (*Arachis hypogaea* L.) crop.

2.3.1.2 Organic Manures and Growth in Groundnut (*Arachis hypogaea* L.)

The application of organic manures together with fertilizers increase groundnut crop growth and yield through improvement in physical conditions of the soil coupled with enhanced supply of nutrients (Ismail et al., 1998; Dosani et al., 1999; Rao and Shaktawat, 2001).

Salama et al. (1994) studied flower production of groundnut under various concentrations of organic manure and water amount.

Taufiq and Sudaryono (1998) observed that chlorotic symptoms on groundnut leaves commonly occurred on high pH alfisol. The symptoms are considered to be caused due to iron deficiency. A survey in the Tuban district showed that the symptoms reduced yield of groundnut planted on alfisol by about 20%. Researches to overcome the symptoms had already been done, but they emphasized management of deficient nutrients. However, the solution was still ambiguous due to the inconsistency of the results. Therefore, this research was aimed at determining alternative management to overcome chlorotic symptoms and increase groundnut productivity on high pH alfisol by sulfur (S) and organic manure application. The research was carried out on alfisol of the Tuban district during the 1996/97 rainy season, using factorial complete block design, replicated three times. Factor I was cow manure application of 0, 5, and 10 t/ha, and the second one was the application of 0, 100, 200, 300, and 400 kg S/ha. The same treatments were tested on the same soil in the Rilet's glasshouse (pot) experiment, using a completely randomized design and replicated five times. Research results indicated that high soil pH (8.4) in alfisol of Tuban was one of the limiting factors for increasing groundnut productivity in the area. However, the low and negative correlation between soil pH and groundnut yield indicated that factors other than soil pH could have also affected the growth and yield of groundnut in these soil. Application of sulfur decreases soil pH and increased soil S content. Organic manure application increased soil organic carbon and was effective as a buffer of soil pH. It can be concluded that application of 200 kg S/ha or 100 kg S/ha combined with 5 t/ha manure increased groundnut yield planted in high pH alfisol. However, the effect of these treatments in alleviating chlorotic symptoms needs more confirmation. It is suggested that the effect of sulfur and organic manure in increasing groundnut yield on high pH alfisol in economic scale needs to be evaluated. It is also important to study further the role of sulfur, gypsum, and ZA on nutrients availability on high pH alfisol.

Bheemaiah et al. (1999) studied the effect of integrated application of green leaf manures and fertilizers on growth and yield of summer groundnut (*Arachis hypogaea*) under different cropping systems.

Fida (2000) set up an experiment in Peshawar in a well-prepared field using randomized complete block design with split plot arrangement having four replications. Soil amendments in the form of percentage of silt were allotted main plots and varieties of groundnut were given subplots. Emergence/m^2 was significantly affected by varieties due to the recommended seed rates for their different growth habits. However, soil amendments and their interaction with varieties had non-significant effect on emergence. Soil amendments and varieties had significant effect on flowering. Maximum days to flowering were taken by the control and semi-erect variety, SP-2000. A number of nodules/plant were statistically affected by varieties but soil amendments had no significant effect to number of nodules/plant. Soil amendments and varieties had a significant effect on nodules weight/plant. Maximum weight of nodules/plant was obtained by 3/4 silt and in spreading varieties like BARD-479. Significant differences were observed in days to maturity due to soil amendments, varieties, and their interaction. Maximum days to maturity were observed in the control and its interaction with spreading variety BARD-479. Maximum fruiting depth was recorded by the erect variety, Parachinar and in soil amendments by 3/4 silt. Maximum pod spreading distance was registered by the interaction of 2/4 silt with spreading variety BARD-479.

The soil amendments, varieties, and their interaction had significantly affected plant height. Tallest plants were noticed in the interaction of erect variety Parachinar and 3/4 silt. Maximum biomass was produced by erect variety Parachinar and in soil amendments by 3/4 silt. Numbers of pods/plant were statistically affected by soil amendments, varieties, and their interaction. Maximum pods/plant were recorded by the interaction of spreading variety BARD-479 with 3/4 silt. Large pods were recorded by the spreading variety BARD-479 and in soil amendments by 3/4 silt. More seeds/pod were attained by the erect variety Parachinar and in soil amendments by 3/4 silt. Maximum thousand seed weight was recorded by the interaction of large-seeded spreading variety BARD-479 and 3/4 silt. Similarly, pod yield was also significantly affected by different soil amendments and varieties. Maximum pod yield was produced by the spreading variety BARD-479 and in soil amendments by 3/4 silt. Finally, soil amendments and varieties had significantly affected oil content. Maximum oil content was registered by the interaction of semi-erect variety SP-2000 and 2/4 silt. Silt at the rate of 3/4 produced maximum yield followed by the non-significant mean of 2/4 silt. In case of varieties, the spreading BARD-479 produced maximum yield followed by the non-significant mean of semi-erect variety Swat Phali-96. Therefore, in this trial, soil amendment of 2/4 silt and the semi-erect variety Swat Phali-96 proved to be the best for Peshawar valley conditions. The reason for recommending the semi-erect variety is that it is easier to harvest than the spreading variety. In case of non-availability of silt, gypsum at the rate of 750 kg/ha could be recommended as well.

Kadam et al. (2000) studied the influence of planting layouts, organic manure, and levels of sulfur on growth and yield of summer groundnut.

Lourduraj (2000) conducted field experiments at the Agricultural Research Station, Aliyar Nagar, Tamil Nadu, India in the summer of 1994 and 1995 on groundnut adopting a split plot design. Two irrigation regimes based on irrigation water/cumulative pan evaporation (IW/CPE) ratios of 0.60 and 0.75 were allotted to the main plot. Twelve treatments involving combinations of organic manure with mineral fertilizer were assigned to the subplots. Irrigation scheduled at IW/CPE 0.75 led to higher plant height, number of branches/plant, dry matter production, and leaf area index, resulting in higher yield than IW/CPE 0.60. All the growth attributes and the yield increased in response to a higher level of mineral fertilizer application and organic manure application. Moreover, combined application of inorganic and organic manures significantly enhanced the growth attributes and yield of groundnut compared to the sole application of either of them.

Rao and Shaktawat (2001) conducted a field experiment in 1997–98 to study the effect of organic manure, phosphorus, and gypsum on growth, yield, and quality of groundnut under rainfed conditions. Application of organic manure significantly increased the number of branches, leaf area index, root dry weight, hydration ratio, and periodic dry matter accumulation (DMA) and thereby yield and quality of groundnut. A mean increase of 14.0% and 11.3% in pod yield was recorded under FYM and poultry manure applications over control (16.22 q ha^{-1}). Application of 60 kg of P_2O_5 ha^{-1} significantly increased growth, yield, and quality parameters compared to 20 kg of P_2O_5 ha^{-1}. The effect of gypsum either in single or in split dose on DMA at 45 DAS was not evident but as crop growth advanced, the effect on all growth parameters was evident. The two modes of gypsum application proved equally effective in increasing pod yield of groundnut.

Talwar et al. (2002) studied the influence of canopy structure and geometry on groundnut productivity; it was examined in two genotypes TMV 2 and TMV 2-NLM. The latter is a mutant of TMV 2 with narrow leaves. The two genotypes were grown on an alfisol field under irrigated and water deficit conditions in the 1994–95 post-rainy and 1995 rainy seasons at ICRISAT. The crop growth rate (CGR) of TMV 2-NLM was greater than TMV 2 under adequately irrigated conditions by 11% in the 1994–95 post-rainy season and by 13% in the 1995 rainy season. Under water deficit conditions, CGR of TMV 2-NLM was 32% higher than in TMV 2. TMV 2-NLM also had greater radiation use efficiency, 0.81 g mj^{-1} compared to 0.68 g mj^{-1} in TMV 2. The light extinction coefficient of TMV 2-NLM was 0.51 as compared to 0.58 of TMV 2 under irrigated conditions, suggesting greater penetration of incident radiation into the canopy of TMV 2-NLM compared to that of TMV 2. Although TMV 2-NLM produced greater total dry matter, the partitioning of dry matter to the pods (P_f) was less compared to TMV 2. Under water deficit conditions, the P_f was reduced by 18% in TMV 2-NLM compared to 13% reduction in TMV 2.

These results suggest scope for enhancing the crop productivity by tailoring canopy architecture. However, further research efforts are required to improve partitioning ability of groundnut genotypes to match enhanced crop growth rates.

Adhikari et al. (2003) conducted a field experiment in the summer seasons of 1996–97, 1997–98, and 1998–99 at the Central Research Farm, Gayeshpur, Nadai, West Bengal, to study the effect of gypsum on growth, yield, and quality of groundnut (*Arachis hypogaea* L.) varieties. The varieties tested were TGS 1, TKG 19A, TG 22, and ICGS 49; the levels of gypsum were 0, 200, and 400 kg/ha. ICGS 49 groundnut showed higher kernel weight (70.5 g), shelling (67.6%), pods/plant (4.7), and oil content (47.2%) than the other varieties. All the varieties performed well at 400 kg gypsum level compared with lower levels of gypsum. However, the highest pod yield (2.38 tons/ha) and net returns (Rs. 35 301/ha) were with 400 kg gypsum/ha. There was no significant difference in yield among TKG 19A, TG 22, and TGS 1. Oil content increased with the increase in gypsum level.

Maity et al. (2003) conducted field experiments at IARI, New Delhi during the kharif (summer) seasons of 1999 and 2000. The experiments revealed that growth parameters like leaf area index (LAI), net assimi-lation rate (NAR), total dry matter (TDM) accumulation, and per plant productivity of groundnut were favored by one month delay in planting of sunflower in groundnut–sunflower intercropping system in replacement series. The NAR of groundnut was decreased at early growth stage due to simultaneous sowing of both the crops, but it was recovered at a later stage and the TDM and productivity were compensated; whereas, late planting resulted in low LAI, TDM, and productivity of sunflower. Total produc-tivity was not influenced by intercropping. In the intercropping system, LAI, TDM, as well as pod/seed weight were increased due to application of phosphorus up to 40 and 80 kg P_2O_5 ha^{-1} in groundnut and sunflower, respectively, while application of sulfur up to 30 kg S ha^{-1} increased these parameters in both the years. However, total productivity was significantly increased due to 40 kg P_2O_5 and 30 kg S ha^{-1}.

Panwar and Singh (2003) conducted field experiments in the rainy season of 2000 and 2001, to study the effect of the conjunctive use of phosphorus and bio-organics on groundnut (*Arachis hypogaea* L.). Appli-cation of 60 kg P_2O_5/ha markedly influenced the growth and yield attri-butes resulting in significant increase of pod yield by 14.01% over the control. Seed inoculation with Rhizobium or phosphorus-solubilizing microorganism (PSM) marginally improved yield, but their combined use increased pod yield significantly. Both the organics, i.e., FYM and neem cake, significantly increased the pod and haulm yield but when half quan-tity of these organics were integrated with Rhizobium and PSM, the high-est pod yield of 31.80 q/ha with FYM 5 tons/ha + Rhizobium + PSM was obtained. Neem cake @ 1.5 tons/ha in the presence of the same biofertil-izers was next to it.

Singh et al. (2003) conducted a field experiment in the summer seasons of 1994 and 1995 at RCA, Udaipur on sandy loam soil to study the response of summer peanut (*Arachis hypogaea* L.) to varying irrigation schedules, organic manures (FYM and CBS) and sulfur nutrition. The irrigation schedule I_3 (0.7/0.7/0.7) and I_2 (0.4/0.7/0.7) based on IW/CPE ratio, being at par, significantly enhanced growth characteristics of summer peanut over irrigation schedule I_1 (0.4/0.7/0.7) in both the years. FYM and CBS were found to be significantly at par in increasing growth parameters over no manure. S at 60 kg ha^{-1} significantly enhanced growth characters of summer peanut while compared to 30 kg S ha^{-1} and no sulfur, respectively, in both the years.

Mohapatra and Dixit (2010) conducted a field experiment at Berhampur in the kharif (summer) season of 2003 and 2004 on loamy sandy soil to study the effect of integrated use of FYM. The recommended dose of fertilizer (RDF) was 20–17.4–33.3 kg NPK/ha Rhizobium, gypsum (250 kg/ha), and boron (1 kg/ha) for good performance of groundnut (*Arachis hypogaea* L.) and soil fertility. Results revealed that application of FYM + 75% RDF + Rhizobium + gypsum + boron recorded significantly higher pods/plant, 100 pod weight resulting in higher pod (2.66 tons/ha), kernel yield (1.92 tons/ha), and nutrient uptake (151.4–17.0–58.6–0.8–0.07 kg NPKSB/ha). This treatment also generated maximum net return (Rs. 22,000/ha). The increase in pod yield of this treatment was 65 and 14% higher over RDF and application of 5 ton FYM + 75% RDF + Rhizobium + lime @ 1/4 lime requirement (LR) + boron, respectively. Application of FYM + 75% RDF + gypsum was the second best pod-yielding treatment. However, the calcium uptake was highest (48.6 kg/ha) with FYM + 75% RDF + Rhizobium + Lime @ 1/4 LR + boron application. There was buildup of available N in the soil but PKSB (phosphorus, potassium, sulfur and boron) were depleted after cropping. Available calcium had depleted in all the treatments except the treatments using lime and combination of FYM and gypsum.

2.3.2 ORGANIC MANURES AND LEAF NUTRIENT (NPK) CONTENTS IN GROUNDNUT (ARACHIS HYPOGAEA L.)

Soon after the establishment of the superiority of shoot analysis over soil analysis as an index of the mineral status of the soil and the mineral requirements of the plant, it was realized that the heterogeneity of the various organs involved, with their diverse functions, and varying quantities of nutrients in each, impaired the credibility of the method. On the other hand, the leaf provided a more uniform organ; leaf analysis proved more dependable for the purpose than the analysis of the entire plants. Lagatu and Maume (1930; 1934) developed the technique of leaf analysis for studying the ions of the absorption of nutrient elements and the influence of different factors that controlled growth with particular reference to the added nutrients (Steward, 1963).

Leaf analysis is now well established as a tool for assessing the nutritional requirements of a particular crop plant (Lundegaradh, 1951). It signifies a definite relationship between the content of the nutrients in the leaves and vegetative growth; the index values of various nutrients at different stages of a crop plant exhibiting a clear picture for the assessment of nutritional requirements. On the other hand, the amount or concentration of various nutrients present in the soil does not always show as a direct correlation with plant growth, as many factors, such as the area of absorbing surface, antagonism and synergism, soil fixation, etc. influence their availability to the physiology of the plant itself; this is beside the several edaphic factors that may also play an important role in the process of absorption.

Reddy (2004) reported that for every 1.0 ton of pods and 2.0 tons of haulms, about 63 N, 11 P_2O_5, 46 K_2O, 27 CaO, and 14 MgO kg ha^{-1} are removed by the groundnut crop. Of these, about 50% nitrogen, 80–90% phosphorous, potassium, calcium, and magnesium are found in haulms. On sandy loam soils, for the production of 100 kg pods, 4.38 kg N, 0.92 P_2O_5, 3.12 kg K_2O, 1.25 kg Mg, and 4.0 g Zn are required. A balanced fertilizer program with particular emphasis on P, K, Ca, and Mg is essential for optimum pod yield. Nitrogen deficiency is characterized by a general chlorosis of the leaves, which become light yellow or nearly white in some cases. The stems become distinctly red besides there being poor nodulation. Lack of phosphorous leads to dark blue-green leaves with reduced size. Characteristic symptoms of chlorosis are the deep red or purple coloration of the stem in later stages and the yellowing and dropping of older leaves. Potassium deficiency leads to stunted growth with drying up of leaf margins. The stem gets a reddish color at the branches. Calcium deficiency

is characterized by development of localized pitted areas on the lower surface of leaves. Later on, large necrotic spots are found on both the leaf
surfaces, which give the leaf a bronze color. The youngest foliage presents
a distorted appearance. The basal stem cracks; in the later stage, dieback of
the shoots occurs. Calcium deficiency leads to poor filling of pods (pops).
Sulfur deficiency leads to pale green color leaves. Magnesium deficiency
leads to interveinal chlorosis, which starts, from the leaf margins and
advances towards the midrib. Iron deficiency produces stunted growth and
characteristic chlorosis symptoms. Small and distorted terminal chlorotic
leaflets with a few yellowish white spots indicate iron deficiency. Boron
deficient leaves resemble calcium deficient leaves except that in the for-
mer, necrotic areas are localized near the leaf margins instead of being
distributed over the entire leaf surface. The primary function of plant analysis is to diagnose problems or to monitor the nutrient status during crop-
growing season for timely correction of problems, if any. Concentrations
of N, P, K, Cu, Mn, and Zn in groundnut leaves, generally, decrease with
increasing age. The concentration of Ca markedly increases with leaf age;
Mg concentration also tends to increase.

Dutta and Mondal (2006) studied the response of groundnut (*Arachis
hypogaea* L.) cv TAG 24 to moisture stress and application of organic
manure (FYM) and fertilizer with and without gypsum in the summer
seasons of 2002 and 2003 at Regional Research Station, BCKV, Jhar-
gram (West Bengal) under acid lateritic soils. No moisture stress as well
as stress at vegetative stage recorded significantly higher values of NPK
uptake than stress at flowering stage. Increase in supply of irrigation provided adequate moisture in the soil, which plays an important role in nutrient uptake involving diffusion, mass flow, and interception. Application of
FYM also influenced markedly the uptake of nutrients over no FYM application. The increase in uptake of nutrients might be attributed to increase
in availability of nutrients due to decomposition of organic compounds on
the application of FYM. The increased P availability was due to the combined effect of released organic acids and organic anions on decomposition of FYM in acid lateritic soils. Inorganic fertilizers when applied with
gypsum increased the uptake of nutrients by the crop. The highest uptake
of nutrients was obtained under the application of 125% RDF along with
gypsum; however, this treatment was on par with that of 100% RDF with

gypsum. It could be attributed to the favorable influence of Ca and S on uptake of NPK in groundnut.

Singh et al. (2007) conducted a field experiment at Wadura in Jammu & Kashmir in the kharif (summer) season of 2004 and 2005 to study the efficiency of Rhizobium, Azotobacter, phosphate-solubilizing bacteria (PSB), and farmyard manure (FYM) on the performance of rainfed soybean (*Glycine max* (L.) Merr.) and soil fertility. Application of Rhizobium + Azotobacter + PSB + FYM (T_{16}) being on par with that of 100% RDF (T_{17}), recorded significantly higher uptake of N and P in both the years over the rest of the treatments. Single inoculation of Azotobacter failed to improve the uptake of N and P, but sole inoculation of soybean with Rhizobium, PSB, and FYM markedly improved their uptake. However, dual as well as multi-inoculation of biofertilizers with or without FYM statistically increased the uptake of N and P.

In the Mohapatra and Dixit experiment described earlier (2010) highest uptake of nutrients (151.4–17–58.6–10.8–0.07 kg NPKSB/ha) were recorded with integrated application of FYM @ 5 tons/ha + 75% recommended dose of fertilizer + Rhizobium + gypsum + B. The combined application of inorganic fertilizer with farm and manure, Rhizobium and B could stimulate the uptake of nutrients due to enhanced microbial and Rhizobium activity, better root growth under the congenial soil physical conditions created by farmyard manure. Increase in P uptake was due to increase in P availability from applied fertilizer and inherent soil source and the combined effect of released organic acids and organic anions on decomposition of farmyard manure in acid lateritic soil. Split application of gypsum ensured adequate Ca and S availability in the fruiting zone at the pegging and pod development stages, which was absorbed by gynophores and increased their uptake. Calcium uptake was highest (48.6 kg/ha) with farmyard manure + 75% recommended dose of fertilizer + Rhizobium + lime + B over other nutrient management practices followed by farmyard manure + 75% recommended dose of fertilizer + Rhizobium + gypsum + B (47.2 kg/ha). The higher Ca uptake due to lime and gypsum application was due to increase in its availability in the rhizosphere. Uptake of B was more due to high demand of boron for pod filling and increase in availability of boron in the root zone.

2.3.3   ORGANIC MANURES AND YIELD IN GROUNDNUT (ARACHIS HYPOGAEA L.)

The flower production of groundnut under various concentrations of organic manure and water amount are studied by Salama et al. (1994).

Ray et al. (1997) anticipated the problem of production constraints in groundnut in Nepal, and started a NARC/ICRISAT collaborative outreach research in 1991. In 1996, farm trials on groundnut were conducted at five locations of five districts. Response of gypsum, pressmud, and lime on groundnut used for confectionery were conducted at two locations of two districts. In a production demonstration trial, three treatments were tested at each location. The mean pod yield of all the locations revealed that there was a 48% pod yield increase due to a high input practice while the yield was increased only by 22% with low input practice over farmer's practice. In response to the application of gypsum, pressmud, and lime trial on groundnut, the mean pod yield in two locations showed that yield increased by 43.7%, 25%, and 18%, respectively.

The Taufiq and Sudaryono experiment (1998), as explained earlier, aimed to find an alternative management practice to overcome chlorotic symptoms and to increase groundnut productivity on high pH alfisol by sulfur (S) and organic manure application.

Kachot et al. (2001) found that the combined application of FYM @ 20 tons/ha + 100% RDF + *Azotobacter* spp. + *Pseudomonas striata* resulted in significantly high pods and haulms—the highest pod and haulm yields being 47.7 and 36.5% higher over the control—and remained comparable with the application of 100% RDF + *Azotobacter* spp. + *Pseudomonas striata*, FYM @ 10 tons/ha + 50% RDF + *Azotobacter* spp. + *Pseudomonas striata*, and 100% RDF in respect to pod and haulm yields. While haulm yield was also found statistically on par with 50% RDF, 50% RDF + *Azotobacter* spp. + *Pseudomonas striata,* and FYM @ 10 tons/ha + 50% RDF. The highest net returns were recorded with the 100% RDF only, which was statistically equal to the combined application of 100% RDF + *Azotobacter* spp. + *Pseudomonas striata*, FYM @ 20 tons/ha + 100% RDF + *Azotobacter* spp. + *Pseudomonas striata*, FYM @ 10 tons/ha + 50% RDF + *Azotobacter* spp. + *Pseudomonas striata,* and FYM @ 10 tons/ha + 50% RDF.

Groundnut yield is an output of sequential metamorphosis from store to sink. Partitioning of photosynthates in vegetative and reproductive parts goes on simultaneously in the later growth phases. The combined effect of FYM, chemical fertilizers, and biofertilizers played a very important role due to their synergistic effect. Application of FYM increased the supply of easily assimilated major as well as micronutrients to plants, besides mobilizing unavailable nutrients into available form. Moreover, biofertilizers also perform better when soil is well supplied with nutrients, particularly nitrogen and phosphorus.

The experiment conducted by Rao and Shaktawat (2001), explained earlier, showed that the application of organic manure significantly increased the pod and biological yield of groundnut over control during both the years and on a pooled basis. The magnitude of mean increase in pod and biological yield due to application of FYM was 14.0 and 10.8% over control. The corresponding increase due to poultry manure application was 11.3 and 9.5%. The significant increases in yield attributes under organic manure application resulted in additional improvement in pod and haulm yields. Such a conducive effect of organic manure could be attributed to the supply of nutrients through mineralization and improvement of physico-chemical properties of the soil.

Patra et al. (1995) reported that DMA at 50 and 70 DAS and LAI at 70 DAS were three key growth variables positively influencing the pod yield of rainfed groundnut. In view of the aforementioned facts, it is evident that FYM and poultry manure application increases crop growth and yield through improvement in physical conditions of soil coupled with enhanced supply of nutrients.

Adhikari et al. (2003) observed, in the experiment explained earlier, that plant height increased with increasing dose of gypsum from 0 to 400 kg/ha. Highest plant height was observed in TG 22, followed by ICGS 49. Height in TKG 19A was significantly lower than that of other varieties. Number of branches did not differ significantly with gypsum application. Number of pod/plant was higher in ICGS 49 and lower in TGS 1. The pods/plant significantly increased with increase in gypsum levels. Higher number of kernels/pod was recorded in ICGS 49. Kernels/pod were highest under 400 kg/ha gypsum, followed 200 kg/ha gypsum. Higher shelling percent was obtained in ICGS 49 and it increased with increasing levels of gypsum, due the fact that gypsum applied at the early flowering stage reduced the number of empty pods. The 100-kernel weight was the highest in ICGS 49 with gypsum 400 kg/ha. The haulm yield was the highest in ICGS 49 and no significant difference was observed in TGS 1 and TKG 19A. Haulm yield increased with increasing gypsum levels. Highest pod yield was obtained in ICGS 49, followed by TG 22. The pod yield increased significantly with increasing gypsum level from 0 to 400 kg. ICGS 49 gave the maximum pod yield when it was treated with gypsum 400 kg/ha. Thus, all the varieties, as mentioned earlier, gave higher yield when they were treated with 400 kg/ha of gypsum. This might be owing to increase in 100-kernel weight and shelling (%).

The Maity et al. (2003) experiment found LAI of groundnuts increased gradually with successive growth stages till harvest. Groundnut in the intercropping system recorded lower LAI than that of sole crop because the intercropping system had only 2/3 population of sole stand. Amongst the intercropping systems, groundnut with staggered planting of sunflower showed higher LAI at all the stages. Sunflower sown simultaneously $(GN + SF_{simul.})$ with groundnut registered higher LAI than the crop sown one month later $(GN + SF_{stagg.})$. The growth of late sown sunflower was suppressed by groundnut crop at all the stages. Groundnut in sole stand showed the highest net assimilation rate (NAR) in the period 30–60 DAS, while groundnut with simultaneously sown sunflower intercrop recorded the lowest during the same period. Since the net photosynthetic rate in groundnut remains high in the upper first to fourth leaves, shading caused by sunflower might be the reason for the lowest NAR in groundnut with simultaneously sown sunflower at early stage of growth. Advancement of sunflower crop towards maturity reduced the LAI and permitted more penetration of solar radiation to reach the groundnut crop as was reflected with the highest NAR value of groundnut in simultaneously sown sunflower intercropping system in the 60–90 DAS period. In case of sunflower, higher NAR was recorded by late sown crop in 30–60 DAS and by normal sown crop from 60 DAS to harvest stage. These two growth periods coincided for both the sunflower crops. Higher NAR of sunflower intercrops during this period might be attributed to prevailing favorable ambient temperature regime. Daily mean temperature during this period ranged between 27.0°C and 29.4°C.

In the Panwar and Singh (2003) experiment, it was noted that plant height, branches, and leaves/plant were significantly improved with the application of biofertilizers combined with organics. The increase in height was possibly due to the elongation of internodes. FYM @ 10 tons/ha and FYM @ 5 tons/ha + Rhizobium (RB) + phosphorus-solubilizing microorganisms (PSM) gave the maximum plant height of 51.34 cm and 51.15 cm, respectively. Highest number of branches and leaves/plant were recorded with FYM @ 5 tons/ha + RM + PSM, followed by neem cake 1.5 tons/ha + RB + PSM. This increase was attributed to the continuous supply of nutrients due to the action of the biofertilizer and release of nutrients from organics. Maximum values of yield attributes, viz., pods/plant, 100-kernel weight, and pod yield/plants, were recorded with FYM 5 tons/ha + RB + PSM. The better response in yield attributes significantly increased the pod yield of groundnut, as there is positive correlation of pod yield with growth and yield attributes. Thus, more branches and leaves/plant increased the number of pods and kernel weight. More pod yield/plant showed its additive effect on pod yield (q/ha) with the application of Rhizobium and PSM combined with FYM @ 5 tons/ha, recording a pod yield of 31.80 q/ha, which was at par with the yield recorded due to neem cake @ 1.5 tons/ha + RB + PSM and significantly superior to the rest of the treatments. The increase in pod yield owing to these treatments may be due to the fact that N and P play an important role in the synthesis of chlorophyll and amino acid, and Rhizobium and PSM ensure the continuous supply of these nutrients, while organics (FYM/neem cake) besides supplying N, P, and K also improved the soil condition, which enhanced the root proliferation and source–sink relationship.

As explained earlier, Reddy (2004) determined that groundnut kernel yield is the product of pod number, number of kernels per pod, and weight of individual kernel. Kernels per pod vary from 2 to 6, pods per plant from 50 to 105, and 100-kernel weight from 28 to 62 g (ICRISAT 1987). This variation is related to cultivar, spacing, fertilizer, and climate. Yield components are most sensitive to environmental stress during flowering and kernel growth stages. There is a high positive correlation between the number of mature pods and pod yield. In general, pod number per plant decrease with increase in plant density. Wider spacing increases the number of branches per plant and number of mature pods per plant. Literature of pod yields indicates maximum contribution to pod yield to from the number of pods per plant followed by 100-kernel weight

Dutta and Mondal (2006) found moisture stress at different growth stages of the crop had a significant influence on yield attributes and yield of groundnut. As mentioned earlier, moisture stress at vegetative stage recorded significantly higher yield attributes, yield, and oil content than at flowering stage; however, this treatment was on a par with a no moisture stress treatment. A temporary soil moisture stress at vegetative stage 10–30 DAS by withholding irrigation did not affect the pod yield significantly, but was highly detrimental when it was imposed at the flowering stage. Organic manure with FYM at 7.5 tons/ha also registered yield advantage compared with no FYM treatment. Significant augmentation in pod yield owing to application of FYM could be ascribed to its favorable effect on yield attributes. The beneficial effect of organic manuring might be due to the improvement in the physical condition of the soil as well as increased availability of plant nutrients, particularly in acid lateritic soil of rice fallow. Application of 100% RDF along with 500 kg gypsum/ha increased significantly the yield and yield attributes of groundnut over 100% RDF or 125% RDF alone; however, this fertilizer combination was on par with 125% RDF and gypsum. This result indicated that gypsum along with fertilizer at recommended level brought about a positive effect on pod yield of groundnut. The additional yield obtained with application of gypsum might be due to its contents of Ca and S, which help increase the pod/plant, kernel/pod, shelling (%), and 100-kernel weight as well as oil content, particularly in sulfur-deficient acid lateritic soils of West Bengal.

Singh et al. (2007) found that biofertilizers inoculation, FYM, and RDF significantly improved the pods/plant, seeds/plant, and 100-seed weight over the control. Maximum increase in yield attributes were noted under RDF, followed by co-inoculation of Rhizobium + *Azotobacter* + PSB + FYM, and Rhizobium + PSB + FYM, which were significantly superior to other treatments. The beneficial effect of biofertilizers might be ascribed to biological N_2 fixation by Rhizobium, activation of amino acids for synthesis of carbohydrates, and phosphate solubilization by phosphate-solubilizing bacteria. Co-inoculation of biofertilizers produced heavier seeds, perhaps due to better translocation of photosynthates. The biofertilizers improved the seed straw yields of soybean, and the application of RDF, being on par with that of Rhizobium + *Azotobacter* + PSB + FYM and Rhizobium + PSB + FYM registered markedly higher value of seed and straw yields over the rest of the treatments. This was due to the synergistic effect among Rhizobium, *Azotobacter,* and phosphate-solubilizing bacteria, which was more pronounced in the presence of FYM. The improvement in seed yield over untreated control was 46.6, 42.2, and 38.8% under RDF, Rhizobium + *Azotobacter* + PSB + FYM, and Rhizobium + PSB + FYM respectively. Further, sole inoculation of Rhizobium and phosphate-solubilizing bacteria registered, respectively 19.8 and 11.2% increase in grain yield over uninoculated control. Rhizobium proved superior to phosphate-solubilizing bacteria, indicating the significance of symbiosis in soybean.

Mohapatra and Dixit (2010) reported integrated application of FYM @ 5 tons/ha + 75% recommended dose of fertilizer + Rhizobium + gypsum @ 250 kg/ha + boron @ 1 kg/ha recorded significantly higher pod yield over all other treatments. Adequate and continuous supply of macro- and micronutrients had positive influence on yield attributes, and recorded 138.8, 48.3, 44.8, 65.2, and 32.7% higher number of mature pods/plant, 100-pod weight, 100-kernel weight, pod yield, haulm yield, respectively over recommended dose of fertilizer. Farmyard manure acted as buffer in the soil with low pH (5.4). It improved the physio-chemical condition of the soil, provided favorable environment, stimulated the uptake of nutrients, and increased the yield over the treatments where FYM was not added. Inoculation with Rhizobium improved the nodulation that enhanced N fixation, activation of amino acids for synthesis of carbohydrates, and consequently was expressed in increase in number of pods/plant, 100-kernel weight, and pod yield.

2.3.4 ORGANIC MANURES AND OIL CONTENT IN GROUNDNUT (ARACHIS HYPOGAEA L.)

The Indian Central Oilseeds Committee (ICOC) was established in 1947 to increase oilseed production through ad hoc research. In 1966, Oilseeds Development Council (ODC) replaced ICOC. Further fillip was given to research by the Project for Intensification of Regional Research on Cotton,

Oilseeds and Millets (PIRCOM). However, increased production could not meet the needs of the ever increasing population. Realizing the problem, the All India Coordinated Research Project on Oilseeds (AICRPO) was setup in 1967 to augment oilseed production. The project generated location-specific technologies for improving the productivity and profitability of oilseeds. The Directorate of Oilseeds Research (DOR), Hyderabad is the lead center for improving the production of seven mandate oilseed crops: rapeseed/mustard, sunflower, safflower, sesame, niger, linseed, and castor. Considering the mounting imports of vegetable oils at the cost of foreign exchange, Technology Mission on Oilseed (TMO) was established in 1986. The TMO was a great success and ushered an era of self-sufficiency in oilseed production with a total production of 22 M tons annually. This is popularly known as the Yellow Revolution. Presently, DOR and AICRPO are concerned with dissemination of technology, production, and distribution of breeders seed through FLDs. Oilseed crops provide not only edible oils but produce a number of by-products used in several industries. Instead of exporting direct items like oilseeds, oil, and oilcakes, India should strive to export value-added products. Since vegetable oils and their derivatives are biodegradable and eco-friendly, they find an immense value in the industry. Groundnut accounts for 28% of the total area and 36% of the total oilseeds production in the country. Gujarat ranks first both in area (1.9 M ha) and production (2.6 M tons). Productivity is highest (1630 kg ha^{-1}) in Tamil Nadu followed by Gujarat (1360 kg ha^{-1}). The four states, Gujarat, Andhra Pradesh, Tamil Nadu, and Karnataka, account for 80% of the area and 80.5% of groundnut production in the country (Reddy, 2004).

The field experiment conducted by Kachot et al. (2001) at Junagadh showed that the shelling (%), protein content, protein yield, and oil yield recorded 3.9%, 13.3, 74.0, and 56.8% higher over the control, correspondingly during individual years as well as in pooled results with combined application of FYM @ 20 tons/ha + 100% RDF + *Azotobacter* spp. + *Pseudomonas striata*, which was found at par with 100% RDF + *Azotobacter* spp. + *Pseudomonas striata*, FYM @ 10 tons/ha + 50% RDF + *Azotobacter* spp. + *Pseudomonas striata,* and 100% RDF application only. Oil content was not influenced significantly by different treatments because oil biosynthesis is a complex process. Hence, it is always difficult to modulate its content in plant through management practices. The probable reason for increase in protein content could be increase in nitrogen and phosphorous, as nitrogen is an integral part of protein and phosphorous is a structural element of certain co-enzymes involved in protein synthesis. The increase in yield of protein and oil is mainly due to the cumulative effect of pod yield. Similar results were also observed by Chawale et al. (1995).

Rao and Shaktawat (2001) observed that application of FYM and poultry manure significantly increased oil and protein content of groundnut kernel. On a pooled basis, oil content increased from 44.8% (under control) to 48.5 and 49.3% under FYM and poultry manure applications. Protein content increased from 22.8% (under control) to 24.2 and 23.9% under FYM and poultry manure application. This increase in oil and protein content under organic manure application can be attributed to the availability of all the essential nutrients in organic matter due to its continuous mineralization.

Adhikari et al. (2003) recorded that both kernel and oil yields increased with the increase in gypsum level. The highest oil yield was obtained in ICGS 49. This might be due to higher oil content and kernel yield. Thus, the net and gross returns increased with the increasing level of gypsum in all the varieties. The benefit–cost ratio and net return were the highest at gypsum 400 kg/ha level. Among the varieties, ICGS 49 gave the highest net return and benefit–cost ratio. Thus, for getting higher yield of confectionery groundnut, higher level of gypsum may be applied. ICGS 49 was found to be the best confectionery variety for West Bengal.

Mohapatra and Dixit (2010) observed that the highest oil yield of 16.5% more than RDF was noted in FYM + 75% RDF + Rhizobium + gypsum + Boron followed by FYM + 75% RDF + gypsum.

2.3.5   CONCLUSION OF REVIEW OF LITERATURE

The review of literature clearly indicated that the low level of productivity of groundnut in India can be ascribed to several constraints including soils low in organic matter content and poor in fertility status. The soil needs judicious organic amendments. Organics; N, P, and K nutrients; as well as other secondary and micro-nutrients, improve the soil condition in an eco-friendly manner, which enhanced the root proliferation. The source–sink relationship need thorough study which is not only scanty but also patchy in groundnut crop. The present study of balanced nutrition integrated with organic manures in groundnut (*Arachis hypogaea* L.) is highly justified.

2.4 METHODOLOGY AND PLAN OF WORK

2.4.1 MATERIAL AND METHODS

2.4.1.1 General

To accomplish the aims and objectives, the following studies were conducted in the field in the kharif (summer) season of the years 2008–2010 at the Botanical Garden and Agricultural Farm, G.F. College, Shahjahanpur, Uttar Pradesh. Kaushal G-201 cultivar of groundnut (*Arachis hypogaea* L.) was selected for the purpose. During the course of work two field experiments were carried out. The detail of the experiments performed, and the prevailing conditions are given in the following section.

2.4.1.2 Agro-Climatic Conditions, Soil Characteristics, and Locality

District Shahjahanpur is found at latitude 27°53′ N, longitude 79°4′ E, and at an altitude of 154.53 meter from the sea level (see, Figure 1.1). It has the semi-arid and sub-tropical climate of the Tarai region with hot dry summers and cold winters. The average rainfall in the years 2008–2009 and 2009–2010 was 560 mm and 748.7 mm, respectively. Most of this annual precipitation was obtained during the two months of June and July. The temperature touched 39.1 and 40.2°C during the crop growth period, and occasionally fell to as low as 6.7 and 9.3°C, respectively in the two years. The meteorological data for the period of the present studies, recorded by the Meteorological Observatory, UP Council of Sugarcane Research, Shahjahanpur, UP have been presented in Figure 2.1 (see also, Figures 1.2 and 1.3).

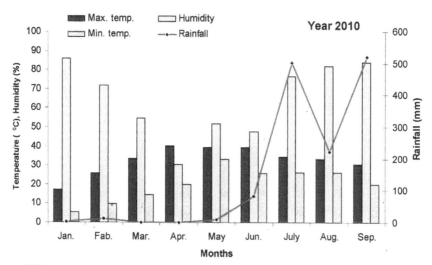

FIGURE 2.1   Showing temperature, relative humidity and rainfall (monthly basis) from January to September 2010.

After preparation of the field, soil samples were collected randomly at a depth of 15 cm from each of the experimental field in the two years and were analyzed for physico-chemical properties. The data are given in Table 1.1.

2.4.2   PREPARATION OF THE FIELD

The field was heavily tilled and leveled before starting each field experiment, the field was thoroughly cleaned to ensure maximum aeration. Microplots of 5.0 sq.m. size were prepared in which uniform recommended basal dose of NPK (N_{50} P_{40} K_{40}) was used similar to as in experiment 1 whereas full and half recommended doses in split design were supplied in experiment 2. Standard agricultural practices recommended for groundnut crop by State Agricultural University Research Institutes were followed.

2.4.3   PROCUREMENT OF SEEDS

Seeds of high quality groundnut (*Arachis hypogaea* L.) variety Kaushal G-201 were procured from CSA University of Agriculture and Technology, Kanpur (UP), India.

2.4.4   PREPARATION OF PRESSMUD COMPOST

The inoculums were prepared by suspending one kg of carrier-based cellu-lolytic culture inoculants (developed by UPCSR, Shahjahanpur) with 100 kg of cattle dung in 500 liters of water for 1 ton of composting material (Singh et al., 2006).

2.4.4.1   Enrichment Inoculum

Nitrogen fixing (Azotobacter) and phosphorus-solubilizing bacteria (PSB) biofertilizers were used for enrichment during the process of composting @ 0.5 kg of each for one ton of composting material.

2.4.4.2   Activators

To accelerate the microbial activities, 8 kg urea + 10 kg single super phos-phate (SSP) were used for each ton of composting material. These are the requirements of microorganisms at the initial stage.

2.4.4.3   Raw Material

Fresh pressmud sulfinated pressmud (SPM) was used as major raw material.

2.4.4.4   Pressmud Composting Process

Pressmud is a soft, spongy, amorphous dark brown to brownish white material containing fiber, sugar, coagulated colloids including cane wax, albuminoids, inorganic salts, and soil particles. The composition varies depending upon the quality of canes crushed and the process followed for the clarification of cane juice in a sugar factory (Yadav, 2001). The range of different constituents of pressmud (sulfitation) is as follows:

S. No.	Constituents	Percent
1	Fiber	15–30
2	Crude wax and fat (lipid)	5–14
3	Crude protein	1–15
4	Total ash	9–20
5	SiO_2	4–10
6	CaO	1–4
7	Mgo	0.5–1.5
8	Total N	1.0–3.1
9	Total P	0.6–3.6
10	Total K	0.3–1.8
11	Sulphur	2–3
12	Organic carbon	35–40

The composting of pressmud was done in a pit. The depth of the pit was 1 m; its width was 1.5 m and length 5 m. This size was sufficient for composting 2 tons of pressmud. The first layer, the bottom most 5 cm, was spread with dry leaves. Weeds, sugarcane trash, or other crop residues was then added followed by a top dressing of mineral fertilizers over it. The second layer consisted of 15 cm of fresh pressmud followed by inoculation with cellulolytic culture inoculants (well suspended with cattle dung in water). This process of spreading two layers was repeated till the height reaches 1 m. The moisture level in the whole process was maintained up to 60%. Finally, the pit was covered with the paste prepared from soil, cattle dung, and weeds. Turnings were given at fortnightly intervals to promote aeration and to maintain optimum moisture and temperature. The pit was covered after each turning.

2.4.4.5 Enrichment

The enrichment with carrier-based N fixing and phosphorous-solubilizing biofertilizers was done at the time of the second turning (after 30 days @ 500 g of each biofertilizers per ton of composting material). The inoculation was done through sprinkling the suspension of biofertilizers in water over the pit.

2.4.4.6   Control

Corresponding control was also maintained in another pit for comparison in which inculcation with cellulolytic culture inoculants and enrichment with biofertilizers was not followed.

2.4.4.7   Maturity

The compost of inoculated pit became ready within 75 days. The compost of both the pits were analyzed for their nutritional and microbial values (Singh et al., 2006).

2.4.5   EXPERIMENT 1

The first experiment was conducted in a factorial randomized design in the kharif (summer) season of 2008–2009. The physico-chemical properties of the soil are given in Table 2.1. This experiment was designed

TABLE 2.1   Physico=chemical analysis of soll of the experimental fields of each experiment conducted during 2008-2010

Characteristics	Experiment 1 (2008–2009)	Experiment 2 (2009–2010)
Txture	Sandy loam	Sandy loam
Particle size distribution		
Sand	70.0	65.0
Silt	15.0	20.0
Clay	15.0	15.0
pH	7.10	7.50
E.C.dS/m	0.30	0.40
Organic Carbon(%)	0.25	0.20
Available N (kg/ha)	196.0	270.00
Available P (kg/ha)	14.10	12.70
Available K (kg/ha)	110.00	115.0
Calcium Carbonate	Nil	Nill

to study the effect of soil-applied organic sources (Pressmud raw with gamma-BHC, pressmud compost, and farmyard manure) at the rate of 0, 5, 10, 15, and 20 tons/ha in groundnut (*Arachis hypogaea* L.) var Kaushal G-201 in terms of growth characteristics, leaf nutrient (NPK) content, yield, and oil content.

A uniform basal dose of 50 kg N, 40 kg P_2O_5, and 40 kg K_2O/ha was applied to each bed before sowing of seeds. All the treatments were replicated thrice. Seeds were treated with thiram before sowing to make them disease free. The row-to-row distance was kept uniformly at 30.0 cm and plant-to-plant distance was 10.0 cm. The sowing was done on July 8, 2009 with treated seeds of groundnut. Organic sources were applied in each bed along with uniform NPK basal dose. The control were supplied with basal NPK dose only. Commercial grade urea, calcium superphosphate, and muriate of potash were used as the respective sources of nitrogen, phosphorus, and potassium. Insecticide (Phosphamidon at 0.075%) was sprayed twice to check aphid infestation, which coincided roughly with the flowering and pegging stage. The crop was irrigated according to standard practices during the entire course of development. Weeding was done when required. The harvesting was done on October 30, 2009. The summary of treatments is given here.

Model of summary of the treatments of Experiment 1 (Factorial Randomize Design)

S No	Treameal	Organic manure(q/ha)		
		Pressmud raw + gama BHC (S₁)	Prewssmud compost (S₂)	Farmyard manure (FYM) (S₃)
1	T₁	0	0	0
2	T₂	5	5	5
3	T₃	10	10	10
4	T₄	15	15	15
5	T₅	20	20	20

2.4.6 EXPERIMENT 2

The second experiment was carried out in the kharif (summer) season of 2009–2010. The physico-chemical properties of the soil are given in Table 2.1. The experiment was of split plot design. The aim of this field trial was to determine the effect of best source and dose of organic manure (i.e., the pressmud compost observed in experiment 1) splitted and applied at different timings (at sowing and flowering) with the split doses of basally applied NPK (full and half) on growth, yield, and oil content of the groundnut. Thus, in a split plot block design, best dose of pressmud compost (20 tons/ha) applied with (at sowing and flowering) splitted, full basal recommended dose of NPK (RDF), i.e., $N_{5050}P_{40}K_{40}$ K/ha and half basal application of NPK, i.e., $N_{25}P_{20}K_{20}$ (at sowing) were tested, to see the economy NPK fertilizer. The summary of the treatment is given below.

Model of summary of the treatments of Experiment 2 (Split Ploat Design)

S. No	Treatment (Manure)	Full basal application of NPK	Half basal application of NPK
1	T_1 = No application	-----	-----
2	T_2 = Full application at sowing	-----	-----
3	T_3 = 1/2 at sowing + 1/2 at flowering	-----	-----
4	T_4 = 2/3 at sowing + 1/3 at flowering	-----	-----
5	T_5 = 1/3 at sowing + 2/3 at flowering	-----	-----
6	T_6 = 1/4 at sowing + 3/4 at flowering	-----	-----
7	T_7 = 3/4 at sowing + 1/4 at flowering	-----	-----

Each treatment was replicated thrice. The size of each plot was 2.0 sq.m. (2.0 × 1.0 m). The crop was sown on June 5, 2010. The seeds were treated with thiram (at 3 g/kg) before sowing to keep them disease free, if any. Sowing was done at 6 cm depth. The crop was seeded with a spacing of 30 × 10 cm. Commercial grade urea, calcium superphosphate, and muriate of potash along with splitted best doses of pressmud compost were applied at the time of sowing. The various plots received irrigations as desired during the course of crop development. The control (T_1) received basal NPK dose full and half RDF only. Weeding was done when required. Harvesting was done on September 29, 2010 when the crop matured.

2.4.7 SAMPLING TECHNIQUES

In experiment 1, three plants from each experimental plot of 2.0×1.0 sq.m were uprooted randomly at various growth stages for the study of various growth characteristics and leaf nutrient (NPK) content of plants. The plants for yield and its attributes with oil content were also randomly collected at harvest only in both the experiments. The yield per hectare was calculated. The pod yield was noted a week after harvest and drying in the sun. Seeds (kernel) were stored for assessing the oil content.

2.4.8 GROWTH CHARACTERS

The germination percentage was noted at a fortnight after sowing. To assess the growth performance, the following parameters were chosen for study at 50, 70, and 90 days of growth in both the experiments.
  i.   Fresh weight (g/plant);
  ii.   Leaf number/plant;
  iii. Plant height (cm);
  iv.   Root length (cm);
  v.   Dry weight (g/plant).
After recording growth characteristics, the leaf sample of each treatment were kept in an oven at 80°C to measure dry weight (till constant weight was achieved). Leaf powder was made and used for leaf NPK content. Chlorophylls and carotenoids were estimated in fresh leaves.

2.4.9 QUANTITATIVE ANALYSIS OF CHLOROPHYLLS AND CAROTENOIDS (ARNON, 1949)

The procedure followed for the quantitative analysis of chlorophylls and carotenoids are similar to that followed in the experiments in Chapter 1 and can be found in Section 1.4.8. Similarly, the procedure for leaf nutrient contents; yield characteristics; procedure for seed oil extraction; and the statistical analysis are also similar and can be found in Sections 1.4.9–1.4.12 (Tables 2.2 and 2.3).

2.5 DATA TABULATION AND ANALYSIS (EXPERIMENTAL RESULTS)

2.5.1 EXPERIMENT 1

In this field experiment, the effect of three soil-applied organic nutrient sources (pressmud raw with gamma-BHC, pressmud compost, and farm-yard manure (FYM) with different rates (0, 5, 10, 15, and 20 g/ha) on germination, growth, leaf nutrient (NPK) content at 50, 70, and 90 days to sowing was studied along with yield characters and oil content at harvest of groundnut variety Kaushal G-201 and of their interaction under local conditions. The data are briefly described in the following sections and summarized in Tables 2.4–2.43.

TABLE 2.2   ANOVA for Organic Source (Experiment 1)

Source of variation	Degree of freedom (d.f.)	Sum of square (SS)	Mean sum of square (MS)	F-value
Replication	2.0			
Source (a)	2.0			
Treatment (b)	4.0			
a x b	8.0			
Error	28.0			
Total	44.0			

TABLE 2.3   ANOVA for Soil-Applied Spplied Splitted Pressmud Compost Application (Experiment 2)

Source of variation	Degree of freedom (d.f.)	Sum of square (SS)	Mean sum of square (MS)	F-value
Replication	2.0			
Source (a)	1.0			
Treatment (b)	6.0			
a x b	6.0			
Error	26.0			
Total	41.0			

2.5.1.1 Germination %

The effect of different soil-applied organic nutrient sources were found to be significant. A gradual significant increase in the number of plant per sq.m was noted as a result of increasing rates of organic manures (Table 2.4) as compared to control supplied with basal RDF only. However, the effect of different sources and their interactions were found to be non-significant (Table 2.4).

2.5.1.2 Growth Characteristics

The effect of soil-applied organic manure rates on different growth characteristics at three growth stages, sources, and their interactions was found to be significant except for the following: the interaction effect of dry weight per plant at 50 days; interaction effect of plant length at all the three stages; chlorophyll a at all the three growth stages; chlorophyll b at 50 days; and carotenoids at all the three growth stages (50, 70, and 90 days)—these were

TABLE 2.4 Effect of Soil-Applied Organic Sources on Germination % in *Arachis hypogaea* L. (Groundnut)

Treatments	Sources (Mean of three replicates)			Mean
	S_1	S_2	S_3	
T_1	93.36	93.10	94.20	93.55
T_2	96.10	96.80	95.50	96.13
T_3	96.30	96.40	97.00	96.57
T_4	96.80	97.20	96.10	96.70
T_5	96.90	98.20	98.00	97.70
Mean	95.89	96.34	96.16	
		S.Em.±	CD at 5%	F-value
Sources		0.33	0.96	NS
Treatments		0.42	1.23	*
Sources x Treatment		0.74	2.14	NS

* = Significant at 5% level of probability;

NS = Non-significant.

TABLE 2.5   Effect of Soil-Applied Organic Sources on Fresh Weight (g/plant) at 50 Days in *Arachis hypogaea* L. (Groundnut)

Treatments	Sources (Mean of three replicates)			Mean
	S_1	S_2	S_3	
T_1	86.00	91.00	86.00	87.67
T_2	89.00	112.00	92.00	97.67
T_3	107.00	127.00	97.00	110.33
T_4	119.00	132.00	113.00	121.33
T_5	127.00	138.00	129.00	131.33
Mean	105.60	120.00	103.40	
		S.Em.±	CD at 5%	F-value
Sources		1.14	3.31	*
Treatments		1.47	4.28	*
Sources x Treatment		2.56	7.41	*

* = Significant at 5% level of probability.

TABLE 2.6   Effect of Soil-Applied Organic Sources on Fresh Weight (g/plant) at 70 Days in *Arachis hypogaea* L. (Groundnut)

Treatments	Sources (mean of three replicates)			Mean
	S_1	S_2	S_3	
T_1	176.60	177.00	181.60	178.40
T_2	226.00	231.30	196.30	217.87
T_3	271.60	312.30	291.60	291.83
T_4	293.30	351.60	336.00	326.97
T_5	427.30	503.00	373.30	434.53
Mean	278.96	315.04	275.76	
		S.Em.±	CD at 5%	F-value
Sources		3.45	10.00	*
Treatments		4.45	12.91	*
Sources x Treatment		7.72	22.37	*

* = Significant at 5% level of probability.

TABLE 2.7   Effect of Soil-Applied Organic Sources on Fresh Weight (g/plant) at 90 Days in *Arachis hypogaea* L. (Groundnut)

Treatments	Sources (Mean of three replicates)			Mean
	S_1	S_2	S_3	
T_1	196.60	194.60	197.00	196.07
T_2	252.30	289.30	288.60	276.73
T_3	314.00	360.30	330.00	334.77
T_4	319.60	395.30	389.60	368.17
T_5	347.30	527.60	391.60	422.17
Mean	285.96.	353.42	319.36	
		S.Em.±	CD at 5%	F-value
Sources		3.26	9.46	*
Treatments		4.21	12.22	*
Sources x Treatment		7.30	21.17	*

* = Significant at 5% level of probability.

TABLE 2.8   Effect of Soil-Applied Organic Sources on Dry Weight (g/plant) at 50 Days in *Arachis hypogaea* L. (Groundnut)

Treatments	Sources (Mean of three replicates)			Mean
	S_1	S_2	S_3	
T_1	17.20	18.20	16.20	17.20
T_2	19.80	22.40	19.30	20.50
T_3	21.40	25.40	19.80	22.20
T_4	23.80	26.40	22.60	24.27
T_5	25.20	27.60	23.80	25.53
Mean	21.48	24.00	20.34	
		S.Em.±	CD at 5%	F-value
Sources		0.33	0.97	*
Treatments		0.43	1.26	*
Sources x Treatment		0.75	2.18	NS

* = Significant at 5% level of probability.

NS = Non-significant.

TABLE 2.9 Effect of Soil-Applied Organic Sources on Dry Weight (g/plant) at 70 Days in *Arachis hypogaea* L. (Groundnut)

Treatments	Sources (Mean of three replicates)			Mean
	S_1	S_2	S_3	
T_1	35.30	36.40	38.30	36.67
T_2	45.20	47.20	42.20	44.87
T_3	54.30	66.40	58.30	59.67
T_4	58.60	70.30	67.20	65.37
T_5	85.40	99.20	78.30	87.63
Mean	55.76	63.90	56.86	
		S.Em.±	CD at 5%	F-value
Sources		0.98	2.85	*
Treatments		1.27	3.68	*
Sources x Treatment		2.20	6.38	*

* = Significant at 5% level of probability.

TABLE 2.10 Effect of Soil-Applied Organic Sources on Dry Weight (g/plant) at 90 Days in *Arachis hypogaea* L. (Groundnut)

Treatments	Sources (Mean of three replicates)			Mean
	S_1	S_2	S_3	
T_1	39.33	40.93	39.40	39.89
T_2	50.46	57.86	49.40	52.57
T_3	62.80	72.06	66.00	66.65
T_4	63.93	79.06	77.93	73.64
T_5	70.73	105.53	74.33	83.53
Mean	57.45	71.09	61.41	
		S.Em.±	CD at 5%	F-value
Sources		1.07	3.11	*
Treatments		1.38	4.02	*
Sources x Treatment		2.40	6.96	*

* = Significant at 5% level of probability.

TABLE 2.11 Effect of Soil-Applied Organic Sources on Leaf Number/Plant at 50 Days in *Arachis hypogaea* L. (Groundnut)

Treatments	Sources (Mean of three replicates)			Mean
	S_1	S_2	S_3	
T_1	40.00	39.80	38.30	39.37
T_2	42.70	52.30	45.60	46.87
T_3	61.80	57.90	50.20	56.63
T_4	72.30	78.10	57.10	69.17
T_5	79.20	82.30	67.40	76.30
Mean	59.20	62.08	51.72	
	S.Em.±	**CD at 5%**	**F-value**	
Sources	1.00	2.92	*	
Treatments	1.30	3.77	*	
Sources x Treatment	2.25	6.53	*	

* = Significant at 5% level of probability.

TABLE 2.12 Effect of Soil-Applied Organic Sources on Leaf Number/Plant at 70 Days in *Arachis hypogaea* L. (Groundnut)

Treatments	Sources (Mean of three replicates)			Mean
	S_1	S_2	S_3	
T_1	84.26	88.16	80.73	84.38
T_2	97.23	99.80	87.30	94.78
T_3	98.40	102.10	93.50	98.00
T_4	107.30	122.30	104.50	111.37
T_5	108.90	123.40	112.00	114.77
Mean	99.22	107.15	95.61	
	S.Em.±	**CD at 5%**	**F-value**	
Sources	1.52	4.42	*	
Treatments	1.97	5.71	*	
Sources x Treatment	3.41	9.89	NS	

* = Significant at 5% level of probability;

NS = Non-significant.

TABLE 2.13   Effect of Soil-Applied Organic Sources on Leaf Number/Plant at 90 Days in *Arachis hypogaea* L. (Groundnut)

Treatments	Sources (Mean of three replicates)			Mean
	S_1	S_2	S_3	
T_1	97.03	110.00	106.00	104.34
T_2	111.60	125.66	117.00	118.09
T_3	128.00	137.00	140.66	135.22
T_4	157.30	177.00	185.66	173.32
T_5	197.66	188.80	186.66	191.04
Mean	138.32	147.69	147.20	
		S.Em.±	CD at 5%	F-value
Sources		2.83	8.20	*
Treatments		3.65	10.59	*
Sources x Treatment		6.33	18.34	NS

* = Significant at 5% level of probability;

NS = Non-signifi-
cant.

TABLE 2.14   Effect of Soil-Applied Organic Sources on Plant Length (cm) at 50 Days in *Arachis hypogaea* L. (Groundnut)

Treatments	Sources (Mean of three replicates)			Mean
	S_1	S_2	S_3	
T_1	55.00	49.30	51.30	51.87
T_2	57.00	56.00	53.00	55.33
T_3	63.30	66.30	63.60	64.40
T_4	69.30	68.60	63.90	67.27
T_5	71.60	68.80	67.60	69.33
Mean	63.24	61.80	59.88	
		S.Em.±	CD at 5%	F-value
Sources		0.70	2.03	*
Treatments		0.90	2.62	*
Sources x Treatment		1.56	4.54	NS

* = Significant at 5% level of probability;

NS = Non-significant.

TABLE 2.15 Effect of Soil-Applied Organic Sources on Plant Length (cm) at 70 Days in *Arachis hypogaea* L. (Groundnut)

Treatments	Sources (Mean of three replicates)			Mean
	S_1	S_2	S_3	
T_1	62.00	67.00	67.30	65.43
T_2	70.30	70.00	71.60	70.63
T_3	70.60	70.90	70.30	70.60
T_4	74.20	73.30	75.30	74.27
T_5	76.60	86.40	79.30	80.77
Mean	70.74	73.52	72.76	
	S.Em.±	CD at 5%	F-value	
Sources	0.81	2.34	NS	
Treatments	1.04	3.03	*	
Sources x Treatment	1.81	5.24	NS	

* = Significant at 5% level of probability;

NS = Non-significant.

TABLE 2.16 Effect of Soil-Applied Organic Sources on Plant Length (cm) at 90 Days in *Arachis hypogaea* L. (Groundnut)

Treatments	Sources (Mean of three replicates)			Mean
	S_1	S_2	S_3	
T_1	66.66	71.66	68.00	68.77
T_2	70.66	73.66	73.66	72.66
T_3	71.90	76.33	74.33	74.19
T_4	75.33	82.00	76.33	77.89
T_5	76.93	88.00	79.90	81.61
Mean	72.30	78.33	74.44	
	S.Em.±	CD at 5%	F-value	
Sources	0.96	2.79	*	
Treatments	1.24	3.60	*	
Sources x Treatment	2.15	6.25	NS	

* = Significant at 5% level of probability;

NS = Non-significant.

TABLE 2.17   Effect of Soil-Applied Organic Sources on Root Length (cm) at 50 Days in *Arachis hypogaea* L. (Groundnut)

Treatments	Sources (Mean of three replicates)			Mean
	S_1	S_2	S_3	
T_1	10.30	11.30	11.35	10.98
T_2	18.60	18.80	18.30	18.57
T_3	18.80	19.50	18.90	19.07
T_4	18.90	19.90	19.10	19.30
T_5	19.10	19.50	19.30	19.30
Mean	17.14	17.80	17.39	
	S.Em.±	**CD at 5%**	**F-value**	
Sources	0.20	0.60	NS	
Treatments	0.27	0.78	*	
Sources x Treatment	0.46	1.35	NS	

* = Significant at 5% level of probability;

NS = Non-significant.

TABLE 2.18   Effect of Soil-Applied Organic Sources on Root Length (cm) at 70 Days in *Arachis hypogaea* L. (Groundnut)

Treatments	Sources (Mean of three replicates)			Mean
	S_1	S_2	S_3	
T_1	19.60	19.60	18.60	19.27
T_2	28.60	31.00	30.30	29.97
T_3	29.60	38.60	36.80	35.00
T_4	40.50	41.00	39.80	40.43
T_5	41.80	52.80	41.60	45.40
Mean	32.02	36.60	33.42	
	S.Em.±	**CD at 5%**	**F-value**	
Sources	0.64	1.87	*	
Treatments	0.83	2.41	*	
Sources x Treatment	1.44	4.19	*	

* = Significant at 5% level of probability.

TABLE 2.19 Effect of Soil-Applied Organic Sources on Root Length (cm) at 90 days in *Arachis hypogaea* L. (groundnut)

Treatments	Sources (Mean of three replicates)			Mean
	S_1	S_2	S_3	
T_1	19.83	19.80	18.83	19.49
T_2	30.16	30.00	29.16	29.77
T_3	29.83	38.00	36.66	34.83
T_4	39.83	40.83	40.33	40.33
T_5	42.83	51.66	40.83	45.11
Mean	32.50	36.06	33.16	
	S.Em.±	CD at 5%	F-value	
Sources	0.49	1.42	*	
Treatments	0.63	1.83	*	
Sources x Treatment	1.09	3.18	*	

* = Significant at 5% level of probability.

TABLE 2.20 Effect of Soil-Applied Organic Sources on Chlorophyll 'a' (mg/g) at 50 Days in *Arachis hypogaea* L. (Groundnut)

Treatments	Sources (Mean of three replicates)			Mean
	S_1	S_2	S_3	
T_1	1.25	1.28	1.28	1.27
T_2	1.28	1.41	1.34	1.34
T_3	1.40	1.56	1.49	1.48
T_4	1.57	1.66	1.56	1.60
T_5	1.59	1.88	1.62	1.70
Mean	1.42	1.56	1.46	
	S.Em.±	CD at 5%	F-value	
Sources	0.029	0.085	*	
Treatments	0.038	0.110	*	
Sources x Treatment	0.066	0.191	NS	

* = Significant at 5% level of probability;

NS = Non-significant.

TABLE 2.21	Effect of Soil-Applied Organic Sources on Chlorophyll 'a' (mg/g) at 70 Days in *Arachis hypogaea* L. (Groundnut)

Treatments	Sources (Mean of three replicates)			Mean
	S_1	S_2	S_3	
T_1	1.44	1.50	1.45	1.46
T_2	1.57	1.82	1.55	1.65
T_3	1.60	1.87	1.57	1.68
T_4	1.72	1.91	1.68	1.77
T_5	1.76	1.93	1.69	1.79
Mean	1.62	1.81	1.59	
		S.Em.±	CD at 5%	F-value
Sources		0.026	0.076	*
Treatments		0.034	0.099	*
Sources x Treatment		0.059	0.171	NS

* = Significant at 5% level of probability;

NS = Non-significant.

TABLE 2.22	Effect of Soil-Applied Organic Sources on Chlorophyll 'a' (mg/g) at 90 Days in *Arachis hypogaea* L. (Groundnut)

Treatments	Sources (Mean of three replicates)			Mean
	S_1	S_2	S_3	
T_1	1.42	1.48	1.51	1.47
T_2	1.49	1.61	1.53	1.54
T_3	1.61	1.62	1.55	1.59
T_4	1.68	1.76	1.62	1.69
T_5	1.83	1.93	1.74	1.83
Mean	1.61	1.68	1.59	
		S.Em.±	CD at 5%	F-value
Sources		0.026	0.076	*
Treatments		0.033	0.098	*
Sources x Treatment		0.058	0.170	NS

* = Significant at 5% level of probability;

NS = Non-significant.

TABLE 2.23   Effect of Soil-Applied Organic Sources on Chlorophyll 'b' (mg/g) at 50 Days in *Arachis hypogaea* L. (Groundnut)

Treatments	Sources (Mean of three replicates)			Mean
	S_1	S_2	S_3	
T_1	0.79	0.83	0.81	0.81
T_2	0.81	0.92	0.87	0.87
T_3	0.86	0.95	0.89	0.90
T_4	0.93	0.99	0.97	0.96
T_5	0.94	1.06	1.02	1.01
Mean	0.87	0.95	0.91	
	S.Em.±	CD at 5%	F-value	
Sources	0.011	0.033	*	
Treatments	0.015	0.043	*	
Sources x Treatment	0.026	0.075	NS	

* = Significant at 5% level of probability;

NS = Non-significant.

TABLE 2.24   Effect of Soil-Applied Organic Sources on Chlorophyll 'b' (mg/g) at 70 Days in *Arachis hypogaea* L. (Groundnut)

Treatments	Sources (Mean of three replicates)			Mean
	S_1	S_2	S_3	
T_1	0.87	0.97	0.85	0.90
T_2	0.95	0.94	0.87	0.92
T_3	0.86	0.98	0.97	0.94
T_4	0.98	1.10	1.01	1.03
T_5	1.03	1.12	1.08	1.08
Mean	0.94	1.02	0.96	
	S.Em.±	CD at 5%	F-value	
Sources	0.011	0.032	*	
Treatments	0.014	0.041	*	
Sources x Treatment	0.025	0.072	*	

* = Significant at 5% level of probability.ccv

TABLE 2.25 Effect of Soil-Applied Organic Sources on Chlorophyll 'b' (mg/g) at 90 Days in *Arachis hypogaea* L. (Groundnut)

Treatments	Sources (Mean of three replicates)			Mean
	S_1	S_2	S_3	
T_1	0.94	1.06	0.96	0.99
T_2	1.01	1.08	0.98	1.02
T_3	1.00	1.07	1.06	1.04
T_4	1.05	1.20	1.02	1.09
T_5	1.04	1.21	1.11	1.12
Mean	1.01	1.12	1.03	
		S.Em.±	CD at 5%	F-value
Sources		0.009	0.026	*
Treatments		0.011	0.034	*
Sources x Treatment		0.020	0.059	*

* = Significant at 5% level of probability.

TABLE 2.26 Effect of Soil-Applied Organic Sources on Carotenoids (mg/g) at 50 Days in *Arachis hypogaea* L. (Groundnut)

Treatments	Sources (Mean of three replicates)			Mean
	S1	S2	S3	
T1	1.04	1.12	1.08	1.08
T2	1.10	1.24	1.29	1.21
T3	1.24	1.34	1.32	1.30
T4	1.31	1.44	1.34	1.36
T5	1.35	1.46	1.34	1.38
Mean	1.21	1.32	1.27	
		S.Em.±	CD at 5%	F-value
Sources		0.014	0.041	*
Treatments		0.018	0.053	*
Sources x Treatment		0.031	0.092	NS

* = Significant at 5% level of probability;

NS = Non-significant.

TABLE 2.27 Effect of Soil-Applied Organic Sources on Carotenoids (mg/g) at 70 Days in *Arachis hypogaea* L. (Groundnut)

Treatments	Sources (Mean of three replicates)			Mean
	S_1	S_2	S_3	
T_1	1.06	1.18	1.11	1.12
T_2	1.16	1.37	1.34	1.29
T_3	1.28	1.49	1.34	1.37
T_4	1.38	1.50	1.45	1.44
T_5	1.43	1.63	1.42	1.49
Mean	1.26	1.43	1.33	
	S.Em.±	CD at 5%	F-value	
Sources	0.014	0.041	*	
Treatments	0.018	0.053	*	
Sources x Treatment	0.032	0.092	NS	

* = Significant at 5% level of probability;

NS = Non-significant.

TABLE 2.28 Effect of Soil-Applied Organic Sources on Carotenoids (mg/g) at 90 Days in *Arachis hypogaea* L. (Groundnut)

Treatments	Sources (Mean of three replicates)			Mean
	S_1	S_2	S_3	
T_1	1.18	1.24	1.21	1.21
T_2	1.24	1.34	1.28	1.29
T_3	1.30	1.36	1.32	1.33
T_4	1.40	1.51	1.39	1.43
T_5	1.44	1.65	1.42	1.50
Mean	1.31	1.42	1.32	
	S.Em.±	CD at 5%	F-value	
Sources	0.014	0.040	*	
Treatments	0.018	0.052	*	
Sources x Treatment	0.031	0.090	NS	

* = Significant at 5% level of probability;

NS = Non-significant.

TABLE 2.29 Effect of Soil-Applied Organic Sources on Leaf Nitrogen (%) at 50 Days in *Arachis hypogaea* L. (Groundnut)

Treatments	Sources (Mean of three replicates)			Mean
	S_1	S_2	S_3	
T_1	2.251	2.249	2.250	2.250
T_2	2.268	2.288	2.270	2.280
T_3	2.433	2.455	2.460	2.450
T_4	2.503	2.520	2.488	2.500
T_5	2.541	2.548	2.540	2.540
Mean	2.400	2.410	2.400	
		S.Em.±	CD at 5%	F-value
Sources		0.027	0.078	NS
Treatments		0.034	0.101	*
Sources x Treatment		0.060	0.175	NS

* = Significant at 5% level of probability;

NS = Non-significant.

TABLE 2.30 Effect of Soil-Applied Organic Sources on Leaf Nitrogen (%) at 70 Days in *Arachis hypogaea* L. (Groundnut)

Treatments	Sources (Mean of three replicates)			Mean
	S_1	S_2	S_3	
T_1	2.272	2.285	2.275	2.280
T_2	2.285	2.325	2.286	2.300
T_3	2.460	2.490	2.470	2.470
T_4	2.525	2.565	2.540	2.540
T_5	2.565	2.570	2.560	2.570
Mean	2.420	2.450	2.430	
		S.Em.±	CD at 5%	F-value
Sources		0.028	0.082	NS
Treatments		0.036	0.106	*
Sources x Treatment		0.063	0.184	NS

* = Significant at 5% level of probability;

NS = Non-significant.

TABLE 2.31 Effect of Soil-Applied Organic Sources on Leaf Nitrogen (%) at 90 Days in *Arachis hypogaea* L. (Groundnut)

Treatments	Sources (Mean of three replicates)			Mean
	S_1	S_2	S_3	
T_1	2.025	2.110	2.100	2.080
T_2	2.115	2.119	2.099	2.110
T_3	2.135	2.210	2.140	2.160
T_4	2.180	2.225	2.190	2.200
T_5	2.210	2.230	2.215	2.220
Mean	2.130	2.180	2.150	
	S.Em.±	**CD at 5%**	**F-value**	
Sources	0.019	0.057	NS	
Treatments	0.025	0.074	*	
Sources x Treatment	0.044	0.129	NS	

* = Significant at 5% level of probability

NS = Non-significant

TABLE 2.32 Effect of Soil-Applied Organic Sources on Leaf Phosphorus (%) at 50 Days in *Arachis hypogaea* L. (Groundnut)

Treatments	Sources (Mean of three replicates)			Mean
	S_1	S_2	S_3	
T_1	0.246	0.248	0.240	0.240
T_2	0.255	0.259	0.250	0.250
T_3	0.260	0.262	0.252	0.260
T_4	0.266	0.270	0.260	0.270
T_5	0.272	0.274	0.270	0.270
Mean	0.260	0.260	0.250	
	S.Em.±	**CD at 5%**	**F-value**	
Sources	0.0038	0.0111	NS	
Treatments	0.0049	0.0144	*	
Sources x Treatment	0.0086	0.0249	NS	

* = Significant at 5% level of probability;

NS = Non-significant.

TABLE 2.33   Effect of Soil-Applied Organic Sources on Leaf Phosphorus (%) at 70 Days in *Arachis hypogaea* L. (Groundnut)

Treatments	Sources (Mean of three replicates)			Mean
	S_1	S_2	S_3	
T_1	0.250	0.256	0.252	0.250
T_2	0.258	0.268	0.259	0.260
T_3	0.266	0.275	0.260	0.270
T_4	0.275	0.280	0.270	0.280
T_5	0.282	0.286	0.276	0.280
Mean	0.270	0.270	0.260	
	S.Em.±	**CD at 5%**	**F-value**	
Sources	0.0033	0.0097	NS	
Treatments	0.0043	0.0126	*	
Sources x Treatment	0.0075	0.0218	NS	

* = Significant at 5% level of probability;

NS = Non-significant.

TABLE 2.34   Effect of Soil-Applied Organic Sources on Leaf Phosphorus (%) at 90 Days in *Arachis hypogaea* L. (Groundnut)

Treatments	Sources (Mean of three replicates)			Mean
	S_1	S_2	S_3	
T_1	0.252	0.260	0.255	0.256
T_2	0.260	0.268	0.262	0.263
T_3	0.265	0.275	0.272	0.271
T_4	0.278	0.280	0.280	0.279
T_5	0.288	0.292	0.286	0.289
Mean	0.269	0.275	0.271	
	S.Em.±	**CD at 5%**	**F-value**	
Sources	0.0025	0.0072	NS	
Treatments	0.0032	0.0093	*	
Sources x Treatment	0.0056	0.0162	NS	

* = Significant at 5% level of probability;

NS = Non-significant.

TABLE 2.35 Effect of Soil-Applied Organic Sources on Leaf Potassium (%) at 50 Days in *Arachis hypogaea* L. (Groundnut)

Treatments	Sources (Mean of three replicates)			Mean
	S_1	S_2	S_3	
T_1	1.055	1.065	1.058	1.059
T_2	1.189	1.200	1.198	1.196
T_3	1.211	1.286	1.210	1.236
T_4	1.219	1.289	1.235	1.248
T_5	1.310	1.322	1.300	1.311
Mean	1.197	1.232	1.200	
	S.Em.±	CD at 5%	F-value	
Sources	0.0087	0.0253	*	
Treatments	0.0113	0.0327	*	
Sources x Treatment	0.0195	0.0567	NS	

* = Significant at 5% level of probability;

NS = Non-significant.

TABLE 2.36 Effect of Soil-Applied Organic Sources on Leaf Potassium (%) at 70 Days in *Arachis hypogaea* L. (Groundnut)

Treatments	Sources (Mean of three replicates)			Mean
	S_1	S_2	S_3	
T_1	1.120	1.135	1.132	1.129
T_2	1.225	1.235	1.215	1.225
T_3	1.260	1.270	1.252	1.261
T_4	1.310	1.333	1.300	1.314
T_5	1.325	1.355	1.340	1.340
Mean	1.248	1.266	1.248	
	S.Em.±	CD at 5%	F-value	
Sources	0.0076	0.0221	NS	
Treatments	0.0098	0.0286	*	
Sources x Treatment	0.0171	0.0495	NS	

* = Significant at 5% level of probability;

NS = Non-significant.

TABLE 2.37   Effect of Soil-Applied Organic Sources on Leaf Potassium (%) at 90 Days in *Arachis hypogaea* L. (Groundnut)

Treatments	Sources (Mean of three replicates)			Mean
	S_1	S_2	S_3	
T_1	1.025	1.011	1.011	1.016
T_2	1.055	1.095	1.040	1.063
T_3	1.086	1.100	1.075	1.087
T_4	1.099	1.120	1.089	1.103
T_5	1.114	1.122	1.120	1.119
Mean	1.076	1.090	1.067	
	S.Em.±	**CD at 5%**	**F-value**	
Sources	0.0061	0.0176	*	
Treatments	0.0078	0.0228	*	
Sources x Treatment	0.0136	0.0395	NS	

* = Significant at 5% level of probability;

NS = Non-significant.

TABLE 2.38   Effect of Soil-Applied Organic Sources on Pod Number/Plant at Harvest in *Arachis hypogaea* L. (Groundnut)

Treatments	Sources (Mean of three replicates)			Mean
	S_1	S_2	S_3	
T_1	16.20	15.90	16.00	16.03
T_2	17.10	18.30	17.10	17.50
T_3	20.20	23.10	20.60	21.30
T_4	25.00	27.20	23.50	25.23
T_5	27.10	28.90	26.90	27.63
Mean	21.12	22.68	20.82	
	S.Em.±	**CD at 5%**	**F-value**	
Sources	0.36	1.05	*	
Treatments	0.46	1.35	*	
Sources x Treatment	0.81	2.35	NS	

* = Significant at 5% level of probability;

NS = Non-significant.

TABLE 2.39 Effect of Soil Applied Organic Sources on Pod Weight (g/plant) at Harvest in *Arachis hypogaea* L. (Groundnut)

Treatments	Sources (Mean of three replicates)			Mean
	S_1	S_2	S_3	
T_1	3.10	3.25	3.15	3.17
T_2	4.00	4.80	3.80	4.20
T_3	5.10	5.60	4.90	5.20
T_4	5.30	5.50	5.60	5.47
T_5	6.00	6.30	5.80	6.03
Mean	4.70	5.09	4.65	
	S.Em.±	CD at 5%	F-value	
Sources	0.052	0.151	*	
Treatments	0.067	0.195	*	
Sources x Treatment	0.116	0.338	*	

* = Significant at 5% level of probability.

TABLE 2.40 Effect of Soil-Applied Organic Sources on Kernel (Seed) Yield (t/ha) at Harvest in *Arachis hypogaea* L. (Groundnut)

Treatments	Sources (Mean of three replicates)			Mean
	S_1	S_2	S_3	
T_1	1.000	1.080	1.100	1.060
T_2	1.200	1.330	1.530	1.353
T_3	1.610	1.980	1.870	1.820
T_4	1.610	2.620	1.980	2.070
T_5	1.680	2.710	2.790	2.393
Mean	1.420	1.944	1.854	
	S.Em.±	CD at 5%	F-value	
Sources	0.025	0.073	*	
Treatments	0.032	0.094	*	
Sources x Treatment	0.056	0.163	*	

* = Significant at 5% level of probability.

TABLE 2.41 Effect of Soil-Applied Organic Sources on Pod Yield (t/ha) at Harvest in *Arachis hypogaea* L. (Groundnut)

Treatments	Sources (Mean of three replicates)			Mean
	S_1	S_2	S_3	
T_1	1.750	1.750	1.780	1.760
T_2	1.960	2.100	1.980	2.013
T_3	2.245	2.260	2.140	2.215
T_4	2.440	2.410	2.490	2.447
T_5	2.663	3.177	3.144	2.995
Mean	2.212	2.339	2.307	
		S.Em.±	CD at 5%	F-value
Sources		0.025	0.073	*
Treatments		0.032	0.295	*
Sources x Treatment		0.057	0.165	*

* = Significant at 5% level of probability.

TABLE 2.42 Effect of Soil Applied Organic Sources on Oil Content (%) at Harvest in *Arachis hypogaea* L. (Groundnut)

Treatments	Sources (Mean of three replicates)			Mean
	S_1	S_2	S_3	
T_1	36.50	36.20	36.40	36.37
T_2	38.00	39.80	40.10	39.30
T_3	39.40	41.20	41.20	40.60
T_4	39.60	41.80	41.80	41.07
T_5	40.00	43.50	43.50	42.33
Mean	38.70	40.50	40.60	
		S.Em.±	CD at 5%	F-value
Sources		0.48	1.41	*
Treatments		0.63	1.83	*
Sources x Treatment		0.10	3.17	NS

* = Significant at 5% level of probability;

NS = Non-significant.

TABLE 2.43   Effect of Soil-Applied Organic Sources on Seed-Oil Yield/(t/ha) at Harvest in *Arachis hypogaea* L. (Groundnut)

Treatments	Sources (Mean of three replicates)			Mean
	S_1	S_2	S_3	
T_1	0.365	0.391	0.400	0.385
T_2	0.456	0.529	0.614	0.533
T_3	0.634	0.816	0.769	0.740
T_4	0.638	1.095	0.836	0.856
T_5	0.672	1.179	1.195	1.015
Mean	0.553	0.802	0.763	
	S.Em.±	CD at 5%	F-value	
Sources	0.008	0.025	*	
Treatments	0.011	0.032	*	
Sources x Treatment	0.019	0.056	*	

* = Significant at 5% level of probability.

noted to be non-significant (Tables 2.26–2.28). The most striking feature observed was that 20 q/ha (T_5) of organic manure proved superior for all the growth characteristics studied irrespective of growth and developmental stage as compared to control supplied with basal RDF only. Regarding various sources, the response to pressmud compost (S_2) was found best with almost all growth parameters. Similarly, regarding various interactions, 20 q/ha of pressmud compost ($T_5 \times S_2$) was found to be significantly the highest in almost all the growth attributes studied. The data are briefly described in the following sections and summarized in Tables 2.5–2.28.

2.5.1.2.1   *Fresh Weight (g/plant)*

A gradual increase in fresh weight of plants was noted as a result of increase in organic manure dose at all the three growth stages and most of the values showed significant differences with each other (Tables 2.5–2.7) as compared to control (T_1). Among the different sources, pressmud compost (S_2) significantly responded best for plant fresh weight at all the three growth stages and the value differed critically with the other organic sources, followed by pressmud raw with gamma-BHC (S_1) statistically equal to farmyard manure (FYM) up to 70 days to sowing but differing critically at 90 days to sowing (Tables 2.5–2.7). The interaction (treatment ×sources) of organic manure for this attribute was found significant at all the three growth stages. Pressmud compost with all the treatments gave comparatively higher values than those of pressmud raw with gamma-BHC and farmyard manure (FYM). However, maximum value was noted in 20 q/ha pressmud compost ($T_5 \times S_2$) as compared to controls (T_1) supplied with basal RDF only (Tables 2.5–2.7).

2.5.1.2.2   Dry Weight (g/plant)

Dry weight per plant was also significantly and gradually increased, similar to fresh weight, as a result of increased dose of organic manure at all the growth stages studied. The highest value was noted with 20 q/ha (T_5) organic manure (Tables 2.8–2.10). The values differed critically with the rest of the treatments as compared to control (T_1) supplied with basal RDF only.

Regarding various sources of organic manure, pressmud compost (S_2) proved best at all the three growth stages for dry weight (g/plant) and the value differed critically with the other sources followed by farmyard manure (S_3), statistically equal to pressmud raw with gamma-BHC (S_1) at all the three growth stages (Tables 2.8–2.10).

As far as the interaction effect was concerned, pressmud compost at the rate of 20 q/ha ($T_5 \times S_2$) showed significantly maximum value for plant dry weight at 70 and 90 days to sowing. However, at 50 days to sowing, the interaction effect for dry weight was found to be non-significant (Table 2.8).

2.5.1.2.3   Leaf Number/Plant

A significant gradual increase in leaf production was noted as a result of increase in the rate of organic manure, maximum being for 20 q/ha (T_5) as compared to control (T_1) supplied with basal RDF only at all the three growth stages. The values differed critically with each other (Tables 2.11–2.13).

The response of pressmud compost (S_2) was best at all growth stages and the value significantly differed with the other sources. However, at

90 days to sowing, the response of pressmud compost (S_2) and farm-yard manure (S_3) for leaf production was noted to be statistically equal (Table 2.13).

At 50 days to sowing, the interaction (organic manure treatment × source) was found best with 20 q/ha pressmud compost ($T_5 \times S_2$). The interaction effect for this parameter at 70 and 90 days to sowing was noted to be non-significant (Tables 2.12–2.13).

2.5.1.2.4   Plant Length (cm)

A significantly gradual increase in plant length, which expresses the vigor, was noted as a result of increase in the amount of organic manure dose—maximum value was noted with 20 q/ha (T_5) at all the three growth stages as compared to control (T_1) supplied with basal RDF only and the values differed critically with each other (Tables 2.14–2.16).

The response of pressmud compost (S_2) was best for this attribute and noted significant at 50 and 90 days to sowing only (Table 2.16) followed by farmyard manure (S_3).

The interaction effect for plant length was noted to be non-significant at all the three growth stages (Tables 2.14–2.16).

2.5.1.2.5   Root Length (cm)

The root length significantly increased as a result of increasing doses of organic manures, the maximum being recorded in 20 q/ha (T_5) as compared to control (T_1) supplied with basal RDF only at all the three growth stages (Tables 2.17–2.19).

The root extension by pressmud compost (S_2) was noted significantly best at 70 and 90 days to sowing (Tables 2.18–2.19) and the values differed critically with the rest of the organic manure sources. The response to pressmud raw with gamma-BHC (S_1) was noted poorest for this parameter.

The interaction effect was significant at 70 and 90 days to sowing only. The best combination was 20 q/ha pressmud compost ($T_5 \times S_2$) at both the advanced stages of growth and development for root extension (Tables 2.8–2.19). The values differed critically with the rest of the combinations.

2.5.1.2.6   Leaf Chlorophyll a (mg/g)

The effect of different organic manuring treatments on leaf chlorophyll a content was significantly and gradually increased at all the three growth stages studied as compared to control (T_1) supplied with basal RDF only (Tables 2.20–2.22).

The response to pressmud compost (S2) for chlorophyll a content was best at all the three stages and the value differed critically with other sources. The interaction effect, was however, non-significant at the three growth stages (Tables 2.20–2.22).

2.5.1.2.7   Leaf Chlorophyll b (mg/g)

Leaf chlorophyll b content was significantly increased as a result of increasing rates of organic manures, the maximum being noted in 20 q/ha (T_5) as compared to control (T_1) supplied with basal RDF only at all the three growth stages studied (Tables 2.23–2.25).

The performance of pressmud compost (S_2) was found significantly best and the value differed statistically with other sources. The performance of pressmud raw with gamma-BHC was poorest irrespective of the stage of development (Tables 2.23–2.25).

The interaction effect, however, for leaf chlorophyll b content was noted significant only at 70 and 90 days to sowing. Among all the combinations, 20 q/ha of pressmud compost ($T_5 \times S_2$) was found best at both the growth stages and the value was statistically equal to 15 q/ha pressmud compost ($T_4 \times S_2$) at the two advanced stages (Tables 2.24 and 2.25).

2.5.1.2.8   Leaf Carotenoid (mg/g)

Leaf carotenoid content was significantly increased due to increase in soil-applied organic manures, highest being recorded in 20 q/ha (T_5) as compared to control (T_1) supplied with basal RDF only at all the three stages of growth and development (Tables 2.26–2.28).

Considering the sources response, pressmud compost (S_2) was again superior for this parameter at all stages of growth. The value differed critically with the rest of the sources (Tables 2.26–2.28). However, the interaction effect of this growth parameter was noted to be non-significant at all the three stages of growth studied (Tables 2.26–2.28).

2.5.1.3   Leaf Nutrient (NPK) Content

Leaf nutrient (NPK) content (%) was significantly affected with an increasing trend as a result of increased amount of organic sources application; the highest values were generally found in 20 q/ha (T_5) as compared to control (T_1) supplied with basal RDF only at all growth stages (Tables 2.29–2.31). The highest values were statistically equal to q/ha (T_4) organic manures for most of the nutrients at almost all growth stages (Tables 2.29–2.37).

The response to pressmud compost (S_2) was significant however only for leaf potassium content at 50 and 90 days to sowing. The interaction effect for leaf nitrogen, phosphorus, and potassium was observed to be non-significant at all growth stages (Tables 2.29–2.37).

2.5.1.3.1   Leaf Nitrogen (%)

Leaf nitrogen (%) was significantly increased due to increased supply of organic manures at all growth stages, maximum being found in 20 q/ha (T_5) as compared to control (T_1) supplied with basal RDF only (Tables 2.29–2.31). However, the value was statistically equal to 15 and 10 q/ha (T_4 and T_3), respectively (Tables 2.29–2.31). The response of different organic manures as well as their interaction for leaf nitrogen (%) were noted to be non-significant (Tables 2.29–2.31).

2.5.1.3.2   Leaf Phosphorus (%)

Leaf phosphorus (%) was gradually and significantly increased due to increase in the amount of organic manure application; the maximum was found in 20 q/ha (T_5), which was statistically equal to 15 and 10 q/ha (T_4 and T_3), respectively except at 90 days where the value differed critically with the rest of the treatments (Tables 2.32–2.34) as compared to control (T_1) supplied with basal RDF only.

The response of different sources as well as their interaction effects at all growth stages were noted to be non-significant (Tables 2.32–2.34) for leaf phosphorus (%).

2.5.1.3.3   Leaf Potassium (%)

Leaf potassium (%) were significantly increased due to increase in organic manure application. The highest value recorded was in 20 q/ha (T_5), followed by 15 q/ha (T_4) at all growth stages as compared to control (T_1) supplied with basal RDF only (Tables 2.35–2.37).

The response of pressmud compost was significantly best at 50 and 90 days to sowing as compared to other sources, followed by farmyard manure (FYM) (Tables 2.35–2.37).

The interaction effects for leaf potassium (%) were noted to be non-significant at all three growth stages studied (Tables 2.35–2.37).

2.5.1.4   Yield Characteristics

At harvest, the effect of different rates of organic manure treatments on various yield characteristics, response of sources, and their interactions were found significant except the interaction effect in the case of pod number and oil content (Tables 2.38–2.43). The important results are described in the following subsections.

2.5.1.4.1   Number of Pods/Plant

A gradual significant increase was found as a result of organic manure treatments; the highest value was noted with 20 q/ha (T_5) and the value differed critically with the rest of the treatments (Table 2.38).

Regarding the organic manure sources, the best response for pod number was shown for pressmud compost (S_2), and it differed critically with that of other sources. The interaction effect for this parameter was, however, noted be non-significant (Table 2.38).

2.5.1.4.2 Pod Weight (g/plant)

As for number of pods, the pod weight also gradually increased as a result of increasing dose of organic manures as compared to control (T_1) supplied with basal RDF only (Table 2.39).

As far as the organic manure response was concerned, maximum pod weight was noted in pressmud compost (S_2) followed by pressmud raw with gamma-BHC (S_1) and farmyard manure (S_3), which gave statistically equal values between each other (Table 2.39).

Among the various interactions, the highest pod weight was noted in 20 q/ha pressmud compost ($T_5 \times S_2$).

2.5.1.4.3 Kernel (Seed) Yield (t/ha)

As for pod weight, the kernel (seed) yield gradually increased as a result of increasing rate of organic manures, the maximum being found in 20 q/ha (T_5) as compared to control (T_1) supplied with basal RDF only. All the values differed critically with each other (Table 2.40).

As far as the response of different sources was concerned, pressmud compost (S_2) was found best followed by farmyard manure (FYM). Pressmud raw with gamma-BHC gave significantly the lowest value for kernel (seed) yield.

Regarding interaction effect, it was clearly found that farmyard manure (FYM) @ 20 q/ha ($T_5 \times S_3$) gave highest value. FYM in general, proved better for all the doses in groundnut kernel (seed) yield and the value was statistically equal to 20 q/ha pressmud compost ($T_5 \times S_2$) combination (Table 2.40).

2.5.1.4.4 Pod Yield (t/ha)

The effect of different treatments of soil-applied organic manure application on pod yield at harvest was noted to be significant. The highest pod yield was noted in 20 q/ha (T_5) treatment (Table 2.41) as compared to control (T_1) supplied with RDF only. The values given by various other soil-application on pod yield differed critically with each other.

Among the different organic manure sources, the response to pressmud compost (S_2) produced the maximum pod yield but the value was equal to that of farmyard manure (FYM) application (S_3). Pressmud raw with gamma-BHC (S_1) responded least and differed critically with other sources (Table 2.41).

Regarding the interaction effect, the pressmud compost at the rate of 20 q/ha ($T_5 \times S_2$) combination was best, statistically equal to farmyard manure supplied at the same rate ($T_5 \times S_3$) for this parameter.

2.5.1.4.5. Oil Content (%)

A gradual significant increase in seed oil content (%) was obtained; the maximum value being found in 20 q/ha (T_5). This value was statistically equal to 15 q/ha (T_4) as compared to control (T_1) supplied with basal RDF only (Table 2.42).

Considering the different sources, response of farmyard manure FYM application (S_3) was maximum but the value was statistically equal to that of pressmud compost (S_2). The interaction effect for oil content (%) was noted to be non-significant (Table 2.42).

2.5.1.4.6   Seed Oil Yield (t/ha)

As for as the seed oil yield at harvest was concerned, a pattern almost similar to pod yield was observed. The maximum seed oil yield was found in 20 q/ha (T_5) application—about 62% more as compared to control (T_1) supplied with basal RDF only (Table 2.43).

Among the different sources, pressmud compost significantly produced 4.8% more seed oil yield as compared to farmyard manure (FYM) source (S_3). The lowest response for seed oil yield was given by pressmud raw with gamma-BHC (S_1).

Regarding the interaction effect, a trend almost similar to kernel (seed) yield and pod yield was noted for seed oil yield too. The maximum value was obtained for the combination of 20 q/ha application in the form of farmyard manure (FYM) interaction ($T_5 \times S_3$) but the value was statistically equal to that of the soil-applied 20 q/ha in the form of pressmud compost ($T_5 \times S_2$) combination (Table 2.43).

2.5.2 EXPERIMENT 2

This field trial was conducted in the rainy season (late kharif) of 2010. It was based on the earlier preliminary experiment carried out by the author herself. The aim of the experiment was to elaborate comprehensively the effect of best organic manure source (pressmud compost @ 20 q/ha) integration by splitting and changing the timing of application under half and full basal RDF. Their interaction effects were noted in split plot design in order to get maximum benefit in connection with fertilizer economy of groundnut (*Arachis hypogaea* L.) crop under local condition. This was done by analyzing germination (%), growth characteristics, leaf nutrient (NPK) content at three growth (50, 70, and 90 days to sowing) stages, and yield characteristics including pod and oil yield at harvest. The important results in brief are considered in the following subsections (Tables 2.44–2.83).

2.5.2.1 Germination (%)

All the treatments gave significantly higher values over control. The maximum germination (%) was recorded in 20 q/ha pressmud full soil-applied at sowing (T_2). The value was statistically equal to 3/4 at sowing + 1/4 at flowering (T_7) as compared to control (T_1) supplied with basal RDF only (Table 2.44). The effects of full and half basal RDF as well as that of interactions (treatments × basal RDF) was noted to be non-significant (Table 2.44).

2.5.2.2 Growth Characteristics

The effect of various treatments, basal RDF, and their interaction effects (treatment × basal RDF) were noted for various growth characteristics. The effect of 20 q/ha soil-applied pressmud, differently splitted amount supplied at sowing and flowering stages ($T_2 × T_7$) showed significant effect recorded at 50, 70, and 90 days to sowing on plant fresh and dry weights, leaf number, plant length, root length, leaf chlorophyll a, b, and

TABLE 2.44   Effect of Different Methods of Soil-Applied Pressmud Compost Application on Germination (%) in *Arachis hypogaea* L. (Groundnut)

Treatments	Soil application basal RDF		Mean (Mean of three replicates)
	Full	Half	
T_1	93.55	93.20	93.38
T_2	95.10	95.20	95.15
T_3	93.60	94.10	93.85
T_4	94.10	94.80	94.45
T_5	94.40	94.20	94.30
T_6	93.60	93.40	93.50
T_7	95.80	94.30	95.05
Mean	94.31	94.17	
	S.Em.±	CD at 5%	F-value
Basal RDF	0.22	0.66	NS
Treatments	0.42	1.24	*
Basal RDF x Treatment	0.60	1.75	NS

* = Significant at 5% level of probability; NS = Non-significant.

TABLE 2.45   Effect of Different Methods of Soil-Applied Pressmud Compost Application on Fresh Weight (g/plant) at 50 Days in *Arachis hypogaea* L. (Groundnut)

Treatments	Soil application basal RDF		Mean (Mean of three replicates)
	Full	Half	
T_1	84.90	80.00	82.45
T_2	135.60	129.30	132.45
T_3	120.00	108.90	114.45
T_4	125.30	120.20	122.75
T_5	110.30	106.20	108.25
T_6	110.10	105.30	107.70
T_7	127.90	120.80	124.35
Mean	116.30	110.10	
	S.Em.±	CD at 5%	F-value
Basal RDF	1.68	4.88	*
Treatments	3.14	9.14	*
Basal RDF x Treatment	4.44	12.92	NS

* = Significant at 5% level of probability; NS = Non-significant.

TABLE 2.46   Effect of Different Methods of Soil-Applied Pressmud Compost Application on Fresh Weight (g/plant) at 70 Days in *Arachis hypogaea* L. (Groundnut)

Treatments	Soil application basal RDF		Mean (Mean of three replicates)
	Full	Half	
T_1	180.00	165.00	172.50
T_2	499.30	415.30	457.30
T_3	271.30	270.00	270.65
T_4	295.40	280.00	287.70
T_5	280.40	265.00	272.70
T_6	265.00	258.00	261.50
T_7	490.50	470.00	480.25
Mean	325.99	303.33	
	S.Em.±	CD at 5%	F-value
Basal RDF	3.62	10.53	*
Treatments	6.78	19.71	*
Basal RDF x Treatment	9.59	27.88	*

* = Significant at 5% level of probability.

TABLE 2.47   Effect of Different Methods of Soil-Applied Pressmud Compost Application on Fresh Weight (g/plant) at 90 Days in *Arachis hypogaea* L. (Groundnut)

Treatments	Soil application basal RDF		Mean (Mean of three replicates)
	Full	Half	
T_1	210.00	195.00	202.50
T_2	530.00	505.40	517.70
T_3	298.30	290.00	294.15
T_4	325.60	305.10	315.35
T_5	315.30	302.10	308.70
T_6	305.10	289.90	297.50
T_7	560.00	530.10	545.05
Mean	363.47	345.37	
	S.Em.±	CD at 5%	F-value
Basal RDF	3.88	11.28	*
Treatments	7.26	21.11	*
Basal RDF x Treatment	10.27	29.86	NS

* = Significant at 5% level of probability; NS = Non-significant.

TABLE 2.48 Effect of Different Methods of Soil-Applied Pressmud Compost Application on Dry Weight (g/plant) at 50 Days in *Arachis hypogaea* **L. (Groundnut)**

| Treatments | Soil application basal RDF | | Mean (Mean of three replicates) |
	Full	Half	
T_1	18.90	15.60	17.25
T_2	27.60	24.50	26.05
T_3	22.40	20.20	21.30
T_4	24.10	22.90	23.50
T_5	22.30	21.40	21.85
T_6	21.80	20.80	21.30
T_7	27.10	26.90	27.00
Mean	23.46	21.76	
	S.Em.±	CD at 5%	F-value
Basal RDF	0.20	0.58	*
Treatments	0.37	1.10	*
Basal RDF x Treatment	0.53	1.55	NS

* = Significant at 5% level of probability; NS = Non-significant.

TABLE 2.49 Effect of Different Methods of Soil-Applied Pressmud Compost Application on Dry Weight (g/plant) at 70 Days in *Arachis hypogaea* L. (Groundnut)

| Treatments | Soil application basal RDF | | Mean (Mean of three replicates) |
	Full	Half	
T_1	38.90	35.20	37.05
T_2	102.60	95.80	99.20
T_3	54.30	52.10	53.20
T_4	66.30	65.20	65.75
T_5	62.50	60.30	61.40
T_6	60.80	56.30	58.55
T_7	101.50	99.80	100.65
Mean	69.56	66.39	
	S.Em.±	CD at 5%	F-value
Basal RDF	0.80	2.33	*
Treatments	1.50	4.37	*
Basal RDF x Treatment	2.12	6.18	NS

* = Significant at 5% level of probability; NS = Non-significant.

TABLE 2.50 Effect of Different Methods of Soil-Applied Pressmud Compost Application on Dry Weight (g/plant) at 90 Days in *Arachis hypogaea* L. (Groundnut)

Treatments	Soil application basal RDF		Mean (Mean of three replicates)
	Full	Half	
T_1	42.50	40.30	41.40
T_2	110.20	105.60	107.90
T_3	72.30	70.20	71.25
T_4	76.80	74.30	75.55
T_5	74.30	74.00	74.15
T_6	68.30	68.10	68.20
T_7	106.80	101.30	104.05
Mean	78.74	76.26	
	S.Em.±	CD at 5%	F-value
Basal RDF	0.77	2.25	*
Treatments	1.44	4.21	*
Basal RDF x Treatment	2.04	5.95	NS

* = Significant at 5% level of probability; NS = Non-significant.

TABLE 2.51 Effect of Different Methods of Soil-Applied Pressmud Compost Application on Leaf Number/Plant at 50 Days in *Arachis hypogaea* L. (Groundnut)

Treatments	Soil application basal RDF		Mean (Mean of three replicates)
	Full	Half	
T_1	40.50	35.40	37.95
T_2	83.50	80.40	81.95
T_3	61.40	60.40	60.90
T_4	68.40	64.30	66.35
T_5	58.40	55.60	57.00
T_6	55.30	54.90	55.10
T_7	82.40	81.00	81.70
Mean	64.27	61.71	
	S.Em.±	CD at 5%	F-value
Basal RDF	0.65	1.89	*
Treatments	1.22	3.55	*
Basal RDF x Treatment	1.72	5.02	NS

* = Significant at 5% level of probability; NS = Non-significant.

TABLE 2.52 Effect of Different Methods of Soil-Applied Pressmud Compost Application on Leaf Number/Plant at 70 Days in *Arachis hypogaea* L. (Groundnut)

Treatments	Soil application basal RDF		Mean (Mean of three replicates)
	Full	Half	
T_1	89.90	80.40	85.15
T_2	128.20	120.30	124.25
T_3	99.80	90.50	95.15
T_4	105.10	99.80	102.45
T_5	100.70	95.60	98.15
T_6	90.80	88.20	89.50
T_7	125.60	120.60	123.10
Mean	105.73	99.34	
	S.Em.±	CD at 5%	F-value
Basal RDF	1.29	3.76	*
Treatments	2.42	7.04	*
Basal RDF x Treatment	3.42	9.96	NS

* = Significant at 5% level of probability; NS = Non-significant.

TABLE 2.53 Effect of Different Methods of Soil-Applied Pressmud Compost Application on Leaf Number/Plant at 90 Days in *Arachis hypogaea* L. (Groundnut)

Treatments	Soil application basal RDF		Mean (Mean of three replicates)
	Full	Half	
T_1	112.00	108.00	110.00
T_2	189.90	178.90	184.40
T_3	125.30	123.30	124.30
T_4	145.60	142.30	143.95
T_5	136.70	130.40	133.55
T_6	124.30	123.90	124.10
T_7	188.80	180.80	184.80
Mean	146.09	141.09	
	S.Em.±	CD at 5%	F-value
Basal RDF	1.92	5.59	NS
Treatments	3.60	10.47	*
Basal RDF x Treatment	5.09	14.8	NS

* = Significant at 5% level of probability; NS = Non-significant.

TABLE 2.54 Effect of Different Methods of Soil-Applied Pressmud Compost Application on Plant Length (cm) at 50 Days in *Arachis hypogaea* L. (Groundnut)

Treatments	Soil application basal RDF		Mean (Mean of three replicates)
	Full	Half	
T_1	40.10	38.20	39.15
T_2	68.00	65.10	66.55
T_3	55.00	50.00	52.50
T_4	60.00	56.20	58.10
T_5	54.20	51.30	52.75
T_6	50.00	48.20	49.10
T_7	66.70	64.10	65.40
Mean	56.29	53.30	
	S.Em.±	CD at 5%	F-value
Basal RDF	0.42	1.24	*
Treatments	0.80	2.32	*
Basal RDF x Treatment	1.13	3.28	NS

* = Significant at 5% level of probability; NS = Non-significant.

TABLE 2.55 Effect of Different Methods of Soil-Applied Pressmud Compost Application on Plant Length (cm) at 70 Days in *Arachis hypogaea* L. (Groundnut)

Treatments	Soil application basal RDF		Mean (Mean of three replicates)
	Full	Half	
T_1	68.50	65.10	66.80
T_2	88.90	86.50	87.70
T_3	75.10	74.10	74.60
T_4	79.80	76.20	78.00
T_5	77.60	74.30	75.95
T_6	73.10	70.10	71.60
T_7	86.50	85.20	85.85
Mean	78.50	75.93	
	S.Em.±	CD at 5%	F-value
Basal RDF	0.65	1.89	*
Treatments	1.22	3.55	*
Basal RDF x Treatment	1.72	5.02	NS

* = Significant at 5% level of probability; NS=Non-significant.

TABLE 2.56   Effect of Different Methods of Soil-Applied Pressmud Compost Application on Plant Length (cm) at 90 Days in *Arachis hypogaea* L. (Groundnut)

Treatments	Soil application basal RDF		Mean (Mean of three replicates)
	Full	Half	
T_1	69.10	68.20	68.65
T_2	92.10	90.00	91.05
T_3	70.20	70.10	70.15
T_4	73.50	71.20	72.35
T_5	70.10	69.80	69.95
T_6	69.40	68.50	68.95
T_7	91.80	90.00	90.90
Mean	76.60	75.40	
	S.Em.±	CD at 5%	F-value
Basal RDF	0.51	1.49	NS
Treatments	0.96	2.79	*
Basal RDF x Treatment	1.36	3.95	NS

* = Significant at 5% level of probability; NS = Non-significant.

TABLE 2.57   Effect of Different Methods of Soil-Applied Pressmud Compost Application On Root Length (cm) at 50 Days in *Arachis hypogaea* L. (Groundnut)

Treatments	Soil application basal RDF		Mean (Mean of three replicates)
	Full	Half	
T_1	10.20	10.10	10.15
T_2	19.90	19.40	19.65
T_3	15.60	15.70	15.65
T_4	17.30	17.10	17.20
T_5	16.20	15.60	15.90
T_6	15.20	14.90	15.05
T_7	19.30	18.30	18.80
Mean	16.24	15.87	
	S.Em.±	CD at 5%	F-value
Basal RDF	0.22	0.66	NS
Treatments	0.42	1.24	*
Basal RDF x Treatment	0.60	1.75	NS

* = Significant at 5% level of probability; NS = Non-significant.

TABLE 2.58 Effect of Different Methods of Soil-Applied Pressmud Compost Application on Root Length (cm) at 70 Days in *Arachis hypogaea* L. (Groundnut)

Treatments	Soil application basal RDF		Mean (Mean of three replicates)
	Full	Half	
T_1	24.10	22.10	23.10
T_2	52.10	49.30	50.70
T_3	35.50	35.40	35.45
T_4	40.20	40.10	40.15
T_5	34.90	34.50	34.70
T_6	32.70	30.40	31.55
T_7	51.80	48.20	50.00
Mean	38.76	37.14	
	S.Em.±	CD at 5%	F-value
Basal RDF	0.34	1.01	*
Treatments	0.65	1.89	*
Basal RDF x Treatment	0.92	2.67	NS

* = Significant at 5% level of probability; NS = Non-significant.

TABLE 2.59 Effect of Different Methods of Soil-Applied Pressmud Compost Application on Root Length (cm) at 90 Days in *Arachis hypogaea* L. (Groundnut)

Treatments	Soil application basal RDF		Mean (Mean of three replicates)
	Full	Half	
T_1	21.50	20.20	20.85
T_2	52.40	51.00	51.70
T_3	40.00	39.80	39.90
T_4	45.10	44.20	44.65
T_5	38.10	37.20	37.65
T_6	35.60	35.00	35.30
T_7	51.40	49.90	50.65
Mean	40.59	39.61	
	S.Em.±	CD at 5%	F-value
Basal RDF	0.25	0.73	*
Treatments	0.47	1.37	*
Basal RDF x Treatment	0.66	1.94	NS

* = Significant at 5% level of probability; NS = Non-significant.

TABLE 2.60   Effect of Different Methods of Soil-Applied Pressmud Compost Application on Chlorophyll 'a' (mg/g) at 50 Days in *Arachis hypogaea* L. (Groundnut)

Treatments	Soil application basal RDF		Mean (Mean of three replicates)
	Full	Half	
T_1	1.28	1.24	1.26
T_2	1.84	1.80	1.82
T_3	1.56	1.54	1.55
T_4	1.60	1.58	1.59
T_5	1.60	1.54	1.57
T_6	1.45	1.42	1.44
T_7	1.82	1.80	1.81
Mean	1.59	1.56	
	S.Em.±	CD at 5%	F-value
Basal RDF	0.019	0.055	NS
Treatments	0.035	0.104	*
Basal RDF x Treatment	0.050	0.147	NS

* = Significant at 5% level of probability; NS = Non-significant.

TABLE 2.61   Effect of Different Methods of Soil-Applied Pressmud Compost Application on Chlorophyll a (mg/g) at 70 Days in *Arachis hypogaea* L. (Groundnut)

Treatments	Soil application basal RDF		Mean (Mean of three replicates)
	Full	Half	
T_1	1.52	1.50	1.51
T_2	1.90	1.90	1.90
T_3	1.70	1.68	1.69
T_4	1.76	1.74	1.75
T_5	1.72	1.70	1.71
T_6	1.60	1.58	1.59
T_7	1.88	1.90	1.89
Mean	1.73	1.71	
	S.Em.±	CD at 5%	F-value
Basal RDF	0.020	0.058	NS
Treatments	0.037	0.108	*
Basal RDF x Treatment	0.053	0.154	NS

* = Significant at 5% level of probability; NS = Non-significant.

TABLE 2.62 Effect of Different Methods of Soil-Applied Pressmud Compost Application on Chlorophyll a (mg/g) at 90 Days in *Arachis hypogaea* L. (Groundnut)

Treatments	Soil application basal RDF		Mean (Mean of three replicates)
	Full	Half	
T_1	1.50	1.48	1.49
T_2	1.91	1.90	1.91
T_3	1.61	1.62	1.62
T_4	1.69	1.65	1.67
T_5	1.60	1.58	1.59
T_6	1.58	1.52	1.55
T_7	1.92	1.88	1.90
Mean	1.69	1.66	
	S.Em.±	CD at 5%	F-value
Basal RDF	0.017	0.051	NS
Treatments	0.033	0.096	*
Basal RDF x Treatment	0.046	0.136	NS

* = Significant at 5% level of probability; NS = Non-significant.

TABLE 2.63 Effect of Different Methods of Soil-Applied Pressmud Compost Application on Chlorophyll b (mg/g) at 50 Days in *Arachis hypogaea* L. (Groundnut)

Treatments	Soil application basal RDF		Mean (Mean of three replicates)
	Full	Half	
T_1	0.81	0.80	0.81
T_2	0.98	0.96	0.97
T_3	0.90	0.86	0.88
T_4	0.92	0.88	0.90
T_5	0.90	0.84	0.87
T_6	0.88	0.85	0.87
T_7	0.98	0.86	0.92
Mean	0.91	0.86	
	S.Em.±	CD at 5%	F-value
Basal RDF	0.0097	0.0282	*
Treatments	0.0181	0.0527	*
Basal RDF x Treatment	0.0256	0.0746	NS

* = Significant at 5% level of probability; NS = Non-significant.

TABLE 2.64   Effect of Different Methods of Soil-Applied Pressmud Compost Application on Chlorophyll b (mg/g) at 70 Days in *Arachis hypogaea* L. (Groundnut)

Treatments	Soil application basal RDF		Mean (Mean of three replicates)
	Full	Half	
T_1	0.96	0.94	0.95
T_2	1.08	1.06	1.07
T_3	0.98	0.98	0.98
T_4	1.00	0.98	0.99
T_5	0.98	0.96	0.97
T_6	0.96	0.94	0.95
T_7	1.10	1.05	1.08
Mean	1.01	0.99	
	S.Em.±	CD at 5%	F-value
Basal RDF	0.0077	0.0225	NS
Treatments	0.0145	0.0421	*
Basal RDF x Treatment	0.0205	0.0596	NS

* = Significant at 5% level of probability; NS = Non-significant.

TABLE 2.65   Effect of Different Methods of Soil-Applied Pressmud Compost Application on Chlorophyll b (mg/g) at 90 days in *Arachis hypogaea* L. (Groundnut)

Treatments	Soil application basal RDF		Mean (Mean of three replicates)
	Full	Half	
T_1	1.06	1.02	1.04
T_2	1.18	1.16	1.17
T_3	1.02	1.00	1.01
T_4	1.08	1.02	1.05
T_5	1.06	1.06	1.06
T_6	1.06	1.04	1.05
T_7	1.20	1.16	1.18
Mean	1.09	1.07	
	S.Em.±	CD at 5%	F-value
Basal RDF	0.0088	0.0257	*
Treatments	0.0165	0.0482	*
Basal RDF x Treatment	0.0234	0.0682	NS

* = Significant at 5% level of probability; NS = Non-significant.

TABLE 2.66 Effect of Different Methods of Soil-Applied Pressmud Compost Application on Carotenoid Content (mg/g) at 50 Days in *Arachis hypogaea* L. (Groundnut)

Treatments	Soil application basal RDF		Mean (Mean of three replicates)
	Full	Half	
T_1	1.10	1.06	1.08
T_2	1.40	1.38	1.39
T_3	1.25	1.24	1.25
T_4	1.30	1.26	1.28
T_5	1.28	1.26	1.27
T_6	1.20	1.18	1.19
T_7	1.40	1.38	1.39
Mean	1.28	1.25	
	S.Em.±	CD at 5%	F-value
Basal RDF	0.0090	0.0263	NS
Treatments	0.0169	0.0493	*
Basal RDF x Treatment	0.0240	0.0697	NS

* = Significant at 5% level of probability; NS = Non-significant.

TABLE 2.67 Effect of Different Methods of Soil-Applied Pressmud Compost Application on Carotenoid Content (mg/g) at 70 Days in *Arachis hypogaea* L. (Groundnut)

Treatments	Soil application basal RDF		Mean (Mean of three replicates)
	Full	Half	
T_1	1.18	1.10	1.14
T_2	1.60	1.56	1.58
T_3	1.30	1.32	1.31
T_4	1.45	1.44	1.45
T_5	1.40	1.35	1.38
T_6	1.25	1.20	1.23
T_7	1.60	1.56	1.58
Mean	1.40	1.36	
	S.Em.±	CD at 5%	F-value
Basal RDF	0.0094	0.0274	*
Treatments	0.0176	0.0513	*
Basal RDF x Treatment	0.0249	0.0726	NS

* = Significant at 5% level of probability; NS = Non-significant.

TABLE 2.68 Effect of Different Methods of Soil-Applied Pressmud Compost Application on Carotenoid Content (mg/g) at 90 Days in *Arachis hypogaea* L. (Groundnut)

Treatments	Soil application basal RDF		Mean (Mean of three replicates)
	Full	Half	
T_1	1.22	1.20	1.21
T_2	1.60	1.56	1.58
T_3	1.40	1.36	1.38
T_4	1.46	1.40	1.43
T_5	1.45	1.40	1.43
T_6	1.35	1.30	1.33
T_7	1.58	1.58	1.58
Mean	1.44	1.40	
	S.Em.±	CD at 5%	F-value
Basal RDF	0.0091	0.0267	*
Treatments	0.0171	0.0499	*
Basal RDF x Treatment	0.0243	0.0706	NS

* = Significant at 5% level of probability; NS = Non-significant.

TABLE 2.69 Effect of Different Methods of Soil-Applied Pressmud Compost Application on Leaf Nitrogen (%) at 50 Days in *Arachis hypogaea* L. (Groundnut)

Treatments	Soil application basal RDF		Mean (Mean of three replicates)
	Full	Half	
T_1	2.250	2.240	2.250
T_2	2.540	2.530	2.540
T_3	2.348	2.335	2.340
T_4	2.414	2.410	2.410
T_5	2.410	2.390	2.400
T_6	2.290	2.286	2.290
T_7	2.550	2.540	2.550
Mean	2.400	2.390	
	S.Em.±	CD at 5%	F-value
Basal RDF	0.0080	0.0234	NS
Treatments	0.0150	0.0438	*
Basal RDF x Treatment	0.0213	0.0619	NS

* = Significant at 5% level of probability; NS = Non-significant.

TABLE 2.70 Effect of Different Methods of Soil-Applied Pressmud Compost Application on Leaf Nitrogen (%) at 70 Days in *Arachis hypogaea* L. (Groundnut)

Treatments	Soil application basal RDF		Mean(Mean of three replicates)
	Full	Half	
T_1	2.286	2.280	2.283
T_2	2.630	2.610	2.620
T_3	2.360	2.340	2.350
T_4	2.450	2.450	2.450
T_5	2.420	2.425	2.423
T_6	2.310	2.310	2.310
T_7	2.650	2.630	2.640
Mean	2.444	2.435	
	S.Em.±	CD at 5%	F-value
Basal RDF	0.0086	0.0251	NS
Treatments	0.0161	0.0470	*
Basal RDF x Treatment	0.0228	0.0665	NS

* = Significant at 5% level of probability; NS = Non-significant.

TABLE 2.71 Effect of Different Methods of Soil-Applied Pressmud Compost Application on Leaf Nitrogen (%) at 90 Days in *Arachis hypogaea* L. (Groundnut)

Treatments	Soil application basal RDF		Mean (Mean of three replicates)
	Full	Half	
T1	2.280	2.270	2.275
T2	2.640	2.605	2.623
T3	2.350	2.345	2.348
T4	2.440	2.425	2.433
T5	2.410	2.410	2.410
T6	2.305	2.300	2.303
T7	2.655	2.640	2.648
Mean	2.440	2.428	
	S.Em.±	CD at 5%	F-value
Basal RDF	0.0049	0.0144	NS
Treatments	0.0093	0.0271	*
Basal RDF x Treatment	0.0131	0.0383	NS

* = Significant at 5% level of probability; NS = Non-significant.

TABLE 2.72   Effect of Different Methods of Soil-Applied Pressmud Compost Application on Leaf Phosphorus (%) at 50 Days in *Arachis hypogaea* L. (Groundnut)

Treatments	Soil application basal RDF		Mean (Mean of three replicates)
	Full	Half	
T_1	0.250	0.248	0.249
T_2	0.274	0.270	0.272
T_3	0.254	0.252	0.253
T_4	0.258	0.254	0.256
T_5	0.252	0.250	0.251
T_6	0.250	0.248	0.249
T_7	0.272	0.270	0.271
Mean	0.259	0.256	
	S.Em.±	CD at 5%	F-value
Basal RDF	0.0031	0.0092	NS
Treatments	0.0059	0.0173	*
Basal RDF x Treatment	0.0084	0.0245	NS

* = Significant at 5% level of probability; NS = Non-significant.

TABLE 2.73   Effect of Different Methods of soil-Applied Pressmud Compost Application on Leaf Phosphorus (%) at 70 Days in *Arachis hypogaea* L. (Groundnut)

Treatments	Soil application basal RDF		Mean (Mean of three replicates)
	Full	Half	
T_1	0.256	0.252	0.254
T_2	0.280	0.278	0.279
T_3	0.260	0.256	0.258
T_4	0.276	0.270	0.273
T_5	0.266	0.260	0.263
T_6	0.258	0.256	0.257
T_7	0.280	0.276	0.278
Mean	0.268	0.264	
	S.Em.±	CD at 5%	F-value
Basal RDF	0.0033	0.0097	NS
Treatments	0.0062	0.0182	*
Basal RDF x Treatment	0.0088	0.0257	NS

* = Significant at 5% level of probability; NS = Non-significant.

TABLE 2.74 Effect of Different Methods of Soil-Applied Pressmud Compost Application on Leaf Phosphorus (%) at 90 Days in *Arachis hypogaea* L. (Groundnut)

Treatments	Soil application basal RDF		Mean (Mean of three replicates)
	Full	**Half**	
T_1	0.262	0.260	0.261
T_2	0.290	0.286	0.288
T_3	0.276	0.274	0.275
T_4	0.285	0.286	0.286
T_5	0.270	0.720	0.495
T_6	0.266	0.266	0.266
T_7	0.288	0.284	0.286
Mean	0.277	0.339	
	S.Em.±	**CD at 5%**	**F-value**
Basal RDF	0.0051	0.0150	*
Treatments	0.0096	0.0281	*
Basal RDF x Treatment	0.0136	0.0397	*

* = Significant at 5% level of probability; NS = Non-significant.

TABLE 2.75 Effect of Different Methods of Soil-Applied Pressmud Compost Application on Leaf Potassium (%) at 50 Days in *Arachis hypogaea* L. (Groundnut)

Treatments	Soil application basal RDF		Mean (Mean of three replicates)
	Full	**Half**	
T_1	1.100	1.055	1.078
T_2	1.320	1.310	1.315
T_3	1.210	1.210	1.210
T_4	1.280	1.270	1.275
T_5	1.280	1.265	1.273
T_6	1.190	1.160	1.175
T_7	1.315	1.305	1.310
Mean	1.242	1.225	
	S.Em.±	**CD at 5%**	**F-value**
Basal RDF	0.0021	0.0150	*
Treatments	0.0096	0.0281	*
Basal RDF x Treatment	0.0136	0.0397	NS

* = Significant at 5% level of probability; NS = Non-significant.

TABLE 2.76   Effect of Different Methods of Soil-Applied Pressmud Compost Application on Leaf Potassium (%) at 70 Days in *Arachis hypogaea* L. (Groundnut)

Treatments	Soil application basal RDF		Mean (Mean of three replicates)
	Full	Half	
T_1	1.130	1.115	1.123
T_2	1.360	1.340	1.350
T_3	1.220	1.210	1.215
T_4	1.290	1.280	1.285
T_5	1.295	1.282	1.289
T_6	1.199	1.190	1.195
T_7	1.350	1.340	1.345
Mean	1.263	1.251	
	S.Em.±	CD at 5%	F-value
Basal RDF	0.0039	0.0115	*
Treatments	0.0074	0.0216	*
Basal RDF x Treatment	0.0105	0.0306	NS

* = Significant at 5% level of probability; NS = Non-significant.

TABLE 2.77   Effect of Different Methods of Soil-Applied Pressmud Compost Application on Leaf Potassium (%) at 90 Days in *Arachis hypogaea* L. (Groundnut)

Treatments	Soil application basal RDF		Mean (Mean of three replicates)
	Full	Half	
T_1	1.085	1.011	1.048
T_2	1.210	1.200	1.205
T_3	1.110	1.105	1.108
T_4	1.140	1.125	1.133
T_5	1.135	1.125	1.130
T_6	1.100	1.050	1.075
T_7	1.220	1.210	1.215
Mean	1.143	1.118	
	S.Em.±	CD at 5%	F-value
Basal RDF	0.0069	0.0201	*
Treatments	0.0129	0.0376	*
Basal RDF x Treatment	1.0183	0.0532	NS

* = Significant at 5% level of probability; NS = Non-significant.

TABLE 2.78 Effect of Different Methods of Soil-Applied Pressmud Compost Application on Pod Number/Plant at Harvest in *Arachis hypogaea* L. (Groundnut)

Treatments	Soil application basal RDF		Mean (Mean of three replicates)
	Full	Half	
T_1	16.80	14.20	15.50
T_2	29.10	24.50	26.80
T_3	22.40	21.40	21.90
T_4	26.40	22.20	24.30
T_5	23.10	21.30	22.20
T_6	19.80	18.90	19.35
T_7	28.90	26.80	27.85
Mean	23.79	21.33	
	S.Em.±	CD at 5%	F-value
Basal RDF	0.30	0.89	*
Treatments	0.57	1.67	*
Basal RDF x Treatment	0.81	2.36	NS

* = Significant at 5% level of probability; NS = Non-significant.

TABLE 2.79 Effect of Different Methods of Soil-Applied Pressmud Compost Application on Pod Weight (g/plant) at Harvest in *Arachis hypogaea* L. (Groundnut)

Treatments	Soil application basal RDF		Mean (Mean of three replicates)
	Full	Half	
T_1	3.50	3.25	3.38
T_2	6.60	6.05	6.33
T_3	4.80	4.70	4.80
T_4	5.50	5.40	5.45
T_5	4.90	4.80	4.85
T_6	4.70	4.45	4.58
T_7	6.55	6.50	6.53
Mean	5.22	5.08	
	S.Em.±	CD at 5%	F-value
Basal RDF	0.046	0.133	*
Treatments	0.086	0.250	*
Basal RDF x Treatment	0.121	0.354	NS

* = Significant at 5% level of probability; NS = Non-significant.

TABLE 2.80   Effect of Different Methods of Soil-Applied Pressmud Compost Application on Kernel (seed) Yield (t/ha) at Harvest in *Arachis hypogaea* L. (Groundnut)

Treatments	Soil application basal RDF		Mean (Mean of three replicates)
	Full	Half	
T_1	1.120	1.000	1.060
T_2	2.850	2.560	2.705
T_3	2.450	2.340	2.395
T_4	2.510	2.150	2.330
T_5	2.010	1.980	1.995
T_6	1.820	1.670	1.745
T_7	2.890	2.950	2.920
Mean	2.236	2.093	
	S.Em.±	CD at 5%	F-value
Basal RDF	0.022	0.065	*
Treatments	0.042	0.122	*
Basal RDF x Treatment	0.059	0.172	*

* = Significant at 5% level of probability.

TABLE 2.81   Effect of Different Methods of Soil-Applied Pressmud Compost Application on Pod Yield (t/ha) at Harvest in *Arachis hypogaea* L. (Groundnut)

Treatments	Soil application basal RDF		Mean (Mean of three replicates)
	Full	Half	
T_1	1.820	1.550	1.685
T_2	3.240	3.100	3.170
T_3	2.810	2.650	2.730
T_4	2.940	2.760	2.850
T_5	2.410	2.320	2.365
T_6	2.100	2.110	2.105
T_7	3.235	3.015	3.125
Mean	2.651	2.501	
	S.Em.±	CD at 5%	F-value
Basal RDF	0.038	0.111	*
Treatments	0.071	0.207	*
Basal RDF x Treatment	0.101	0.292	NS

* = Significant at 5% level of probability; NS = Non-significant.

TABLE 2.82 Effect of Different Methods of Soil-Applied Pressmud Compost Application on Oil Content (%) in *Arachis hypogaea* L. (Groundnut)

Treatments	Soil application basal RDF		Mean (Mean of three replicates)
	Full	Half	
T_1	36.10	36.00	36.05
T_2	42.50	41.50	42.00
T_3	40.10	39.50	39.80
T_4	40.80	40.30	40.55
T_5	38.90	38.50	38.70
T_6	38.80	38.40	38.60
T_7	41.80	41.20	41.50
Mean	39.86	39.34	
	S.Em.±	CD at 5%	F-value
Basal RDF	0.23	0.69	NS
Treatments	0.44	1.30	*
Basal RDF x Treatment	0.63	1.84	NS

* = Significant at 5% level of probability; NS = Non-significant.

TABLE 2.83 Effect of Different Methods of Soil-Applied Pressmud Compost Application on Seed-Oil Yield (t/ha) at Harvest in *Arachis hypogaea* L. (Groundnut)

Treatments	Soil application basal RDF		Mean (Mean of three replicates)
	Full	Half	
T1	0.404	0.360	0.382
T2	1.208	1.062	1.135
T3	0.982	0.924	0.953
T4	1.024	0.866	0.945
T5	0.782	0.767	0.775
T6	0.706	0.641	0.674
T7	1.208	1.215	1.212
Mean	0.902	0.834	
	S.Em.±	CD at 5%	F-value
Basal RDF	0.0085	0.0249	*
Treatments	0.0160	0.0465	*
Basal RDF x Treatment	0.0226	0.0658	*

* = Significant at 5% level of probability..

total leaf carotenoid content (Tables 2.45–2.68) in (*Arachis hypogaea* L.) (groundnut) crop.

The effect of full and half basal RDF application was also noted significant for the growth characteristics except plant length at 90 days, root length at 50 days, chlorophyll a at all the three stages, chlorophyll b at 70 days, and carotenoid at 50 days to sowing (Tables 2.45–2.68). The interaction effect, however, for all the growth parameters were noted to be non-significant except for fresh weight at 70 days to sowing (Tables 2.46). The most important result of these growth parameters are given in the following subsections.

2.5.2.2.1 Fresh Weight (g/plant)

All the treatments gave significantly higher values over control at the three stages of growth studied for this parameter (Tables 2.45–2.47). At 50 days fresh weight was highest in presumed full supplied at sowing (T_2) statistically equal to ¾ at sowing + 1/4 at flowering (T_7) as compared to control (T_1) supplied with basal RDF only (Table 2.45). The fresh weight at 70 and 90 days to sowing was found maximum in pressmud supplied 3/4 at sowing + 1/4 at flowering (T_7) and the value differed critically with the rest of the treatment (Tables 2.46 and 2.47).

As far as the basal RDF was concerned, full basal dose was best for fresh weight at all the stages (Tables 2.45–2.47). However, the interaction effects (treatment × basal RDF) were noted to be non-significant at 50 and 90 days to sowing. At 70 days, full pressmud supplied at sowing (T_7) × full RDF was best and statistically equal to pressmud supplied 3/4 at sowing + 1/4 at flowering (T_7) × full RDF as well as 3/4 at sowing + 1/4 at flowering (T_7) × half RDF between each other (Table 2.46).

1.5.1.2.2 Dry Weight (g/plant)

Dry weight production as a result of different treatments was noted to be significant at all the three stages of growth (Tables 2.48–2.50). At 50 and 70 days to sowing, pressmud supplied 3/4 at sowing + 1/4 at flowering (T_7) gave highest value, statistically equal to full applied at sowing (T_2).

At 90 days, pressmud supplied full at sowing (T_2) gave highest dry weight statistically equal to pressmud supplied 3/4 at sowing + 1/4 at flowering (T_7) as compared to control (Table 2.50).

Among the split plots, full RDF was found better than half basal RDF at all the three stages. The interaction effect (treatment × basal RDF), however, was noted non-significant at all the three growth stages (Tables 2.48–2.50).

2.5.2.2.3 Leaf Number/Plant

The effect of different timings of splitted pressmud compost application on leaf production was found significant at all the three growth stages (Table 2.51–2.53). The highest leaf number at 50 and 70 days to sowing was noted in full pressmud applied at the time of sowing (T_2) and the value was statistically equal to 3/4 applied at sowing + 1/4 applied at flowering (T_7) as compared to control (T_1) supplied with basal RDF only. At 90 days to sowing, the situation reversed and the highest value was noted in 3/4 applied at sowing + 1/4 applied at flowering (T_7) statistically equal to full pressmud supplied at the time of sowing (T_2) (Table 2.53) as compared to control (T_1).

The response to basal RDF was significant only at 50 and 70 days to sowing, being full RDF as better than half RDF at both the stages (Tables 2.51–2.52). The interaction (treatment × basal RDF) effects were noted to be non-significant at all the three stages for leaf production (Tables 2.51–2.53).

2.5.2.2.4 Plant Length (cm)

The effect of different pressmud splitted dose application on plant length of *Arachis hypogaea* L. (groundnut) was noted significant at all the three growth stages (Tables 2.54–2.56). The highest value was noted in full pressmud applied at the time of sowing (T_2), statistically equal to 3/4 applied at sowing + 1/4 applied at flowering (T_7) as compared to control (T_1) supplied with basal RDF only (Tables 2.54–2.56).

The response to full basal RDF was better at 50 and 70 days to sowing as compared to half basal RDF at the initial two stages for plant height.

The response to basal RDF was, however, non-significant at 90 days to sowing for plant height. The interaction effects (treatments × basal RDF) were however noted to be non-significant at all the three growth stages.

2.5.2.2.5   Root Length (cm)

The effect of different treatments on root length was noted to be significant at all three stages of growth (Tables 2.57–2.59). The highest root length was noted in full pressmud applied at the time of sowing (T_2), statistically equal to 3/4 applied at sowing + 1/4 applied at flowering (T_7) as compared to control (T_1) supplied with basal RDF only.

The response to basal RDF for root length at 50 days was noted to be non-significant. At 70 and 90 days to sowing, full basal RDF was better compared to half basal RDF (Tables 2.58 and 2.59). The interaction effects (treatments × basal RDF) were noted be non-significant at all the three growth stages for this parameter (Tables 2.57–2.59).

2.5.2.2.6   Leaf Chlorophyll a (mg/g)

The effect of different treatments on leaf chlorophyll a content at three growth stages was noted be significant (Tables 2.60–2.62). The maximum value was noted in full pressmud application at the time of sowing (T_2), statistically equal to 3/4 applied at sowing + 1/4 flowering (T_7) as compared to control (T_1) supplied with basal RDF only.

The response to basal RDF as well as the interaction effects (treatments × basal RDF) for leaf chlorophyll a contents were noted to be non-significant at all the three growth stages (Tables 2.60–2.62).

2.5.2.2.7   Leaf Chlorophyll b (mg/g)

The effects of different timings of splitted pressmud compost application in *Arachis hypogaea* L. (groundnut) on leaf chlorophyll b content was noted significant at all the three growth stages (Tables 2.63–2.65). At 50 days to sowing, full pressmud application at the time of sowing (T_2) was best, statistically equal to 3/4 applied at sowing + 1/4 applied at flowering (T_7). However, at 70 and 90 days to sowing, 3/4 pressmud applied at sowing + 1/4 at flowering (T_7) was best, statistically equal to full dose applied at the time of sowing (T_2) as compared to control (T_1) supplied with basal RDF only (Tables 2.63–2.65).

The response to basal RDF was noted significant at 50 and 90 days to sowing. The maximum response was noted for full basal RDF as compared to half basal RDF for leaf chlorophyll b content. The interaction effects (treatments × basal RDF), however, was noted be non-significant at all the three growth stages (Tables 2.63–2.65).

2.5.2.2.8 Leaf Carotenoid Content (mg/g)

Leaf carotenoid contents were significantly affected by different timings of splitted pressmud compost application in *Arachis hypogaea* L. (groundnut) at all the three growth stages (Tables 2.66–2.68).

Maximum value was recorded in full pressmud applied at the time of sowing (T_2), statistically equal to 3/4 applied at sowing + 1/4 applied at flowering (T_7) as compared to control (T_1) supplied with basal RDF only (Table 2.66–2.68).

The response to full basal RDF was better as compared to half basal RDF at 70 and 90 days to sowing (Table 2.67–2.68). The interaction effects (treatment × basal RDF), however were noted to be non-significant for leaf carotenoid content at all the three growth stages (Tables 2.66–2.68).

2.5.2.3 Leaf Nutrient (NPK) Contents

Leaf nutrient (NPK) contents were significantly affected by different timings of splitted pressmud compost application in *Arachis hypogaea* L (groundnut) at all the three growth stages as compared to control (Table 2.69–2.77). The response to basal RDF for leaf phosphorus content at 90 days and at all the three growth stages for leaf potassium contents only were noted significant. The interaction effects (treatments × basal RDF) was noted significant only for leaf phosphorus at 90 days (Table 2.74), considering all of them (Tables 2.69–2.77).

2.5.2.3.1   Leaf Nitrogen (%)

Leaf nitrogen (9%) was significantly affected due to different treatments at all the three growth stages (Tables 2.69–2.71). The maximum leaf nitrogen (%) was noted in pressmud applied 3/4 at sowing + 1/4 at flowering (T_7); it was statistically equal to full pressmud applied at the time of sowing (T_2) as compared to control (T_1) supplied with basal RDF only. The response to basal RDF as well as interaction effects (treatments × basal RDF) for leaf nitrogen at the three growth stages were noted to be non-significant (Tables 2.69–2.71).

2.5.2.3.2   Leaf Phosphorus (%)

The effects of different treatments on leaf phosphorus (%) was significantly affected at all the three growth stages (Tables 2.72 to 2.74); the highest value being recorded in full pressmud applied at the time of sowing (T_2), which was statistically equal to 3/4 applied at sowing + 1/4 applied at flowering (T_7) as compared to control (T_1) supplied with basal RDF only. The response to full basal RDF was only significantly best at 90 days to sowing for leaf phosphorus content. However, the interaction effects (treatments × basal RDF) were noted to be non-significant for leaf phosphorus (%) except at 90 days, where full pressmud applied at the time of sowing with full basal RDF (T_2 × full basal RDF) was best and statistically equal to 3/4 applied at sowing + 1/4 applied at flowering with full basal RDF (T_7 × full basal RDF) combination (Tables 2.74).

2.5.2.3.3   Leaf Potassium (%)

Leaf potassium (%) was significantly affected due to different treatments at all the three growth stages (Tables 2.75–2.77). At 50 and 70 days to sowing, this parameter was found best in full pressmud supplied at the time of sowing (T_2) and was statistically equal to 3/4 applied at sowing + 1/4 applied at flowering (T_7) as compared to control (T_1) supplied with basal RDF only. At 90 days to sowing, the highest value for leaf potassium (%) was recorded in pressmud supplied 3/4 at sowing + 1/4 applied at flowering (T_7), statistically equal to full pressmud supplied at the time of sowing (T_2).

The response to full basal RDF was better as compared to half basal RDF for leaf potassium (%) at all the three growth stages. The interaction effects (treatment × basal RDF) for these nutrient contents were noted to be non-significant at all the three growth stages (Tables 2.75–2.77).

2.5.2.4   Yield Characteristics

The yield characteristics at harvest (pod number and weight, kernel and pod yield as well as oil content and yield) were significantly affected due to different timings of splitted applied doses of pressmud compost in *Arachis hypogaea* L. (groundnut). The response to basal RDF as well as the interaction effects for these yield parameters were also noted significant except for the response to basal RDF for oil content and interaction effect for pod number and weight as well as pod yield and oil content (Tables 2.78–2.83).

2.5.2.4.1   Number of Pods/Plant

The pod number significantly increased due to different treatments (Tables 2.78). The highest number of pods were produced in pressmud supplied 3/4 at sowing + 1/4 applied at flowering (T_7), statistically equal to full pressmud supplied at the time of sowing (T_2) as compared to control (T_1) supplied with basal RDF only. The interaction effect for pod number at harvest was noted to be non-significant (Table 2.78).

The response to full basal RDF for pod number was better than half basal RDF (Table 2.78).

2.5.2.4.2   Pod Weight (g/plant)

At harvest, pod weight was significantly affected by different treatments (Table 2.79). The highest value was recorded for pressmud supplied 3/4 at sowing + 1/4 at flowering (T_7), statistically equal to full pressmud supplied at the time of sowing (T_2) as compared to control (T_1) supplied with basal RDF only. The response to full basal RDF was higher as compared to half basal RDF for pod weight in groundnut. The interaction effects (treatment × basal RDF) was noted to be non-significant for this yield parameters (Table 2.79).

2.5.2.4.3   Kernel (Seed) Yield (t/ha)

Kernel (seed) yield was significantly affected due to different treatments (Table 2.80). The highest kernel (seed) yield was noted in pressmud supplied 3/4 at sowing + 1/4 at flowering (T_7) followed by full pressmud applied at the time of sowing (T_2). Considering the interaction effects (treatment × basal RDF), 3/4 pressmud supplied at sowing + 1/4 supplied at flowering (T_7) with half basal RDF recorded maximum seed yield (T_7 × half basal RDF), statistically equal to (T_7 × full basal RDF) as well as T_2 × full basal RDF combinations (Table 2.80).

2.5.2.4.4   Pod Yield (t/ha)

At harvest, pod yield was significantly affected due to different timings of splitted pressmud compost application in *Arachis hypogaea* L. (groundnut) crop (Table 2.81). The highest pod yield was noted in full pressmud applied at the time of sowing (T_2), statistically equal to pressmud applied 3/4 at sowing + 1/4 at flowering (T_7) as compared to control (T_1) supplied with basal RDF only (Table 2.81).

The response to full basal RDF for pod yield was better as compared to half basal RDF. The interaction effects (treatments × basal RDF), however, were noted to be non-significant for pod yield (Table 2.81).

2.5.2.4.5   Oil Content (%)

The effect of different treatments was noted significant on oil content (%) in *Arachis hypogaea* L. (groundnut) crop (Table 2.82). The highest oil content was found in full pressmud applied at the time of sowing (T_2), statistically equal to 3/4 applied at sowing + 1/4 applied at flowering (T_7) as compared to control (T_1) supplied with basal RDF only. The response to basal RDF as well as interaction effects (treatments × basal RDF) were noted to be non-significant for this yield parameter (Table 2.82).

2.5.2.4.6 Seed Oil Yield (t/ha)

The effect of different timings of splitted pressmud compost on oil yield was found significant at harvest in *Arachis hypogaea* L. (groundnut) crop. Highest oil yield was recorded in 3/4 pressmud applied at sowing + 1/4 applied at flowering (T_7) followed by full pressmud supplied at the time of sowing (T_2) as compared to control (T_1) supplied with basal RDF only (Table 2.83). The response to full basal RDF was better compared to half basal RDF for oil yield. As far as the interaction effects (treatments × basal RDF) were concerned, the combination (T_7 × half) of 3/4 pressmud supplied at sowing + 1/4 at flowering with half basal RDF gave highest oil yield statistically equal to (T_7 × full) as well as full pressmud applied at sowing with full basal RDF (T_2 × full) combination (Table 2.83).

2.6 CONCLUSION: CROP YIELD POTENTIAL

2.6.1 GENERAL

Currently, the desire to manipulate plant morphology, anatomy and function in the interests of crop improvement and environmental protection is intense. The linkages between physiology, morphology, climate, agronomy and yield limits are well known. Field trials are needed to quantify the factors governing current yield barriers and suggest ways of taking crops to their ultimate yield limits in tomorrow's world which would have significant role in predicting the relative importance of yield shaping traits of each crop. The time course of growth in the field represents the integration of numerous biological and environmental processes that conclude with harvestable yield. At its simplest, crop growth can be divided in to two phases; the first concerned largely with vegetative components, and the second with the development of the reproductive organs and seeds growth.

Genetic constitution is primary responsible for determining the pattern of plant growth and yield. Nevertheless, they are also markedly influenced by various other factors including available nutrients, climate, cultural practices, population density, season, and soil. An ideal combination of these factors ensures optimum yield, which has always been the aim of researchers as well as farmers. Soil is the important source of plant nutrients. When the crop requirements are higher than the soil supplying power, nutrients are applied as manures or fertilizer, or both. Manures are plant and animal wastes that are used as sources of plant nutrients. They release nutrients after their decomposition. Manures can be grouped into bulky organic manures and concentrated organic manures based on concentration of the nutrients. Bulky organic manures contain small percentage of nutrients and are applied in large quantities. Farmyard manure (FYM), compost, and green manure are the most important and widely used bulky organic manures. Publications on these lines with regard to groundnut, an important cash and oil crop among the crop plants necessitating periodical researches, include Kanwar et al. (1983); Gillar and Morvan (1984); Venkaiah (1985); Survace et al. (1986); Sims (1986) Golakiya (1988); Manoharan et al. (1988); Agasimani and Hosmani (1989); Salama et al. (1994); Chawale et al. (1995); Patra et al. (1995); Trivedi et al. (1995); Lourduraj et al. (1996); (2000); Ramesh et al. (1997); Anandswarup et al. (1998); Ismail et al. (1998); Taufiq and Sudaryono (1999); Dosani et al. (1999); Kadam et al. (2000); Kachot et al. (2001); Mishra (2001); Rao and Shaktawat (2001); (2002); Panwar and Singh (2003); Bhattacharyya and Chakraborty (2005); Dutta and Mondal (2006); Kumar et al. (2010) and Singh et al. (2010); Latha and Sharma (2013); Chavan et al. (2014). A perusal of these and other publications highlights the importance of studies under balanced fertilization integrated with organic manures regarding: (i) growth characteristics and leaf nutrient (NPK) contents; (ii) yield attributes and oil yield; and (iii) economy of inputs (fertilizers) with respect to final yield of groundnut.

It may not be out of place to mention here that work done at Shahjahanpur, during the last two decades or so on the same important aspects of applied physiology of cereals, sugar, medicinal and oil crops under the supervision of Prof. Abbas, has yielded valuable results: Abbas et al., 1980; Abbas and kumar, 1987; Hasan, 2009; Hasan and Abbas, 2007a, b and 2008; Hasan et al., 2007; 2009; Kanaujia, 2008; Kishor, 2006; Kishor and Abbas, 2003, Kishor et al., 2006 Kumar and Abbas, 1992; Kumar, 2008 Kumar, 2009; Kumar et al., 2007 ; 2008 a; Kumar et al., 2009 a, b; Sharma 2007; Shukla, 2010; Kumar et al., 2010; and Sajjad et al., 2011; Kumar et al., 2014.

It is worthwhile to note here, by the intensive survey of literature on the subject, that balanced nutrition integrated with organic manures in groundnut (*Arachis hypogaea* L.) has not been appropriately dealt with (Verma, 2008), particularly under local conditions to improve sustained pod and oil yield. Therefore, research on this problem has been carried out by the present author by including different rates of soil-applied manures, pressmud raw with gamma-BHC and, pressmud compost in addition to farmyard manure (FYM) (Experiment 1), as well as selecting the best (pressmud compost) manure and dose by varying application method, splitted amount with timing of its application in groundnut under full and half RDF to achieve fertilizer economy. The results obtained in the two field experiments undertaken in 2008–2009 and 2009–2010, laying particular emphasis on pod and oil yield are given here.

2.6.2   EXPERIMENT 1

This field experiment was conducted to study the performance of Kaushal G-201 variety of groundnut (*Arachis hypogaea* L.) under three soil-applied organic manure sources (pressmud raw with gamma-BHC, pressmud compost, and farmyard manure) @ 0, 5, 10, 15, and 20 q/ha and their interaction on growth, yield, and oil content. It was laid out according to a factorial randomized block design by supplying a constant basal dose of 50 kg N, 40 kg P, and 40 kg K/ha before sowing of seeds.

It was observed that increasing soil-applied organic manures doses gradually increased pod and oil yield (Tables 2.41 and 2.43) due to higher values of growth, leaf nutrient content, throughout the growth period (Tables 2.5–2.28). Such results are not uncommon and have been obtained, by other researchers also, such as Kanwar et.al. (1983) and Panwar and Singh (2003) in groundnut. It was also observed that organic manuring improves the physical condition of the soil as well as increased availability

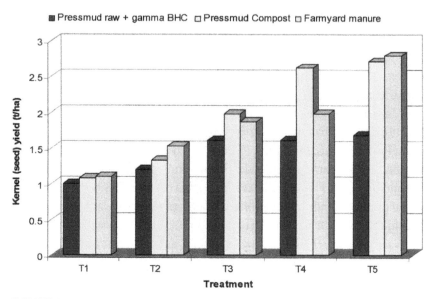

FIGURE 2.2 Effect of soil-applied organic sources on kernel (seed) yield (t/ha) at harvest in *Arachis hypogaea* L. (Experiment 1).

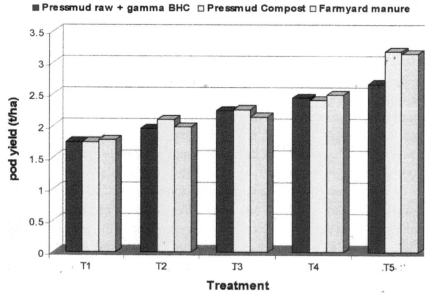

FIGURE 2.3 Effect of soil-applied organic sources on pod yield (t/ha) at harvest in *Arachis hypogaea* L. (Experiment 1).

of plant nutrients as has also been reported by Mishra (2001); Rao and Shaktawat (2001), and Dutta and Mondal (2006).

Significant (41.2%) increase in pod yield and 62.06% oil yield was recorded due to soil-applied 20 q/ha organic manure (Tables 2.41 and 2.43; Figures 2.2 and 2.3), as compared to control supplied with basal RDF only. This increase was associated with the significant increases in yield attributes under organic manure application resulting in additional improvement in pod number and weights as well as oil content percentages (Tables 2.38–2.42). Such a conducive effect of organic manure could be attributed to the supply of nutrients through mineralization as revealed by leaf nutrient (NPK) content at all growth stages showing more uptake of these nutrients (Tables 2.29–2.37). It could also be due to improvement in rhizosphere environment as reported by Deva Kumar and Giri (1998). Among the various interactions, soil-applied 20 q/ha pressmud compost produced highest pod yield (Table 2.42; Figure 2.3), statistically equal to soil-applied 20 q/ha FYM. Total oil yield was found highest by the interaction of soil-applied 20 q/ha FYM (Table 2.43), statistically equal to soil-applied 20 q/ha pressmud compost (Figure 2.4). The highest pod yield

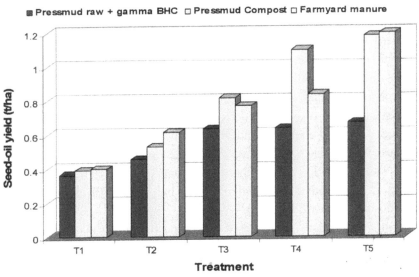

FIGURE 2.4   Effect of soil-applied organic sources on seed-oil yield (t/ha) at harvest in *Arachis hypogaea* L. (Experiment 1)

noted in soil-applied 20 q/ha pressmud compost was associated with better germination (%), fresh and dry weight, leaf number, plant and root length, chlorophylls and carotenoid content, as well as leaf nutrient (NPK) content at almost all growth stages (Tables 2.5–2.37) in this interaction leading to improved yield attributing factors (Figure 2.2; Tables 2.38–2.43) in the pod and oil yield (Tables 2.41 and 2.43). Organic manures acted better in soil with low pH (5.4), improving yield attributing factors (Mohapatra and Dixit, 2010; Table 2.38–2.43) and hence, the pod and oil yield (Figures 2.3 and 2.4). Organic manures also acted as buffer in the soil with low pH (5.4) and improved nutrient availability from organic and inorganic source (Mohapatra and Dixit, 2010). The increase in uptake of leaf (NPK) nutrients due to pressmud compost (Tables 2.29–2.37) by groundnut crop appears to be due to the cumulative effect of increased pod yield. The results substantiated the finding of Chawala et al. (1995); Trivedi et al. (1995); Ramesh et al. (1997); and Kachot et al. (2001) with combined application of chemical fertilizers and organic manures.

2.6.3 EXPERIMENT 2

Organic manures provide a good substrate for the growth of many microorganisms and maintain a favorable nutritional balance and soil physical properties (Singh et al., 2007).

The second field experiment was based on the results of the first experiment done by the author in connection with the effect of different rates of best organic manure source in groundnut on growth, leaf nutrient (NPK) content, pod and oil yield to determine the superior efficacy of the best source (pressmud compost) under local sub-tropical Tarai region of Uttar Pradesh by splitting the optimum amount (20 q/ha) and changing the time of its soil application to get enhanced utilization of soil-applied pressmud compost in groundnut crop under full and half basal NPK dose in a split plot design (Tables 2.44–2.83).

Full soil-application of pressmud compost (organic manure) at sowing produced 46.81% more pod yield as compared to control (basal RDF only), which was statistically equal to 3/4 at sowing plus 1/4 at flowering (Table 2.81), due to improvement in pod number and pod weight (Tables

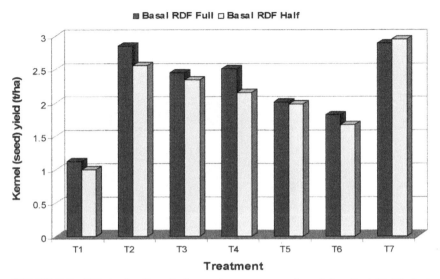

FIGURE 2.5 Effect of soil-applied organic sources on kernel (seed) yield (t/ha) at harvest in *Arachis hypogaea* L. (Experiment 2)

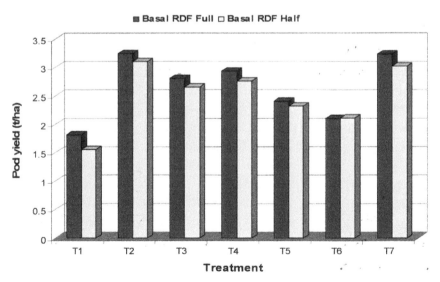

FIGURE 2.6 Effect of different methods of soil-applied pressmud compost application on pod yield (t/ha) at harvest in *Arachis hypogaea* L. (Experiment 2)

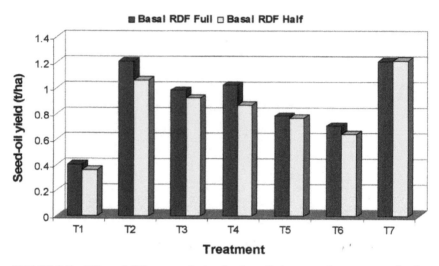

FIGURE 2.7 Effect of different methods of soil-applied pressmud compost application on seed-oil yield (t/ha) at harvest in *Arachis hypogaea* L. (Experiment 2)

2.78 and 2.79) and better growth parameters like germination, fresh and dry weight, leaf number, plant and root length, chlorophylls and carotenoids (Tables 2.44–2.68), and higher leaf nutrient (NPK) content (Tables 2.69–2.77). According to Singh et al. (2010), higher seed yield in groundnut was mainly attributed to greater biomass, pod weight, shelling percent and harvest index, and sustained regulated enrichment of the soil fertility status by combined organic and inorganic treatments (Figures 2.5–2.7). Nonetheless, Singh et al. (2007) also noted significantly improved pods/plant, seeds/plant and 100-seed weight over control due to biofertilizers and organic manure application in soyabean. According to them, biofertilizers and manures increased the uptake of N and P. This might be attributed to enhanced activity of nitrogenase and nitrate reductase enzyme in the soil (Sarawgi et al., 1998), leading to greater biological N_2 fixation (Singh et al., 2007). A positive correlation of pod yield with growth and yield attributes in groundnut crop due to organic manures (Tables 2.44–2.83; Figures 2.5–2.7) has also been noted by Manoharan et al. (1988) and Panwar and Singh (2003).

The combined application of chemical fertilizer along with enough bulk of organic manure has always stimulated the uptake of N (Anandswarup et al., 1998)—partly because of stimulated microbes flush and improved root growth due to congenial soil physical condition created by addition of the heavy bulk of organic manure (Golakiya, 1988; Dutta and Mondal, 2006). The enhanced effects of pressmud compost on growth yield and its attributes, are therefore well understandable in this experiment.

Application of 20 q/ha pressmud compost 3/4 at sowing + 1/4 at flowering resulted in 46.07% more oil yield as compared to control (Figure 2.7), statistically equal to full soil-application at sowing besides supplying N. P and K also provided improved soil condition throughout the growth period and maturity, which enhanced root proliferation (Tables 2.57–2.59) and source–sink relationship (Figure 2.5). The results corroborate the findings of Bhatacharya and Ghosh (2001) and Panwar and Singh (2003). Thus the productivity of groundnut can be further increased considerably by changing the time, split dose of soil-application of pressmud compost organic manure under sub-tropical Tarai condition of Shajahanpur district.

Among the various interaction effects, the highest value of oil yield for half soil-applied RDF was obtained in the case of 20 q/ha of pressmud compost 3/4 at sowing + 1/4 at flowering (Table 2.83). Clearly, there was a saving of 25 kg N, 20 kg P, and 20 kg k/ha (half basal RDF) beside yielding 3.36% more oil as compared to full soil-applied basal RDF by the same treatment (Figure 2.7). Therefore, this combination of soil-applied, half basal RDF and 20 q/ha of pressmud compost 3/4 at sowing + 1/4 at flowering may be recommended for highest profit with fertilizer economy.

2.6.4 PROPOSALS FOR FUTURE WORK

It is evident from the aforementioned discussion that some of the problems in relation to integrated use of mineral nutrition groundnut crop have been solved. However, in view of the low soil fertility status and poor yield of this crop under local conditions, new high-yielding genotypes with biofertilizers intercropped with long duration crops may be tried to exploit the genetic diversity and generate more income and profit.

The effect of various agronomic variables like seeding date and rate in combination with various organic manures and biofertilizers along with amount of irrigation (including water stress) may be studied, also taking into account rainfed conditions.

Further, mere increase in pod and oil yield would be meaningless unless the content of the unsaturated fatty acids, particularly palmitic, oleic, linoleic, linoleinic acids, is also increased. The present work may, therefore, be extended to test the effect of various agro-techniques on oil quality of this crop.

Among the different organic manures, the effect of pressmud compost with basal RDF proved best for high yields. However, the efficiency of bioinoculants like Rhizobium, Azotobacter, phosphate-solubilizing bacteria alone and in combination with gypsum and micronutrients may be tried in future to determine further improvements in pod and oil yield of groundnut crop. In addition, pressmud compost, which gave a very promising performance in the present experiments, may also be further included in future trials. Understanding the yield production process in future will provide the insight needed to design plants that will produce higher yields in the challenging fluctuating global environment.

KEYWORDS

- *Arachis hypogaea* L.
- fertilizer
- FYM
- groundnut
- oil content
- pod yield
- pressmud compost

REFERENCES

Abbas, Z., & Kumar, S. (1987). Response of sonalika wheat and triticale varieties to timings of nitrogen application coinciding with irrigation schedule. *Geobios, 14,* 27–30.

Abbas, Z., Samiullah, Afridi, M. M. R. K., & Inam, A. (1980). Effect of different levels of nitrogen on the fodder yield of five varieties of rainfed sorghum. *Comp. Physiol. Ecol., 5,* 143–145.

Adhikari, J., Samanta, D., & Samui, R. C. (2003). Effect of gypsum on growth and yield of confectionary groundnut (*Arachis hypogaea*) varieties in summer season. *Indian Journal of Agricultural Sciences, 73,* 108–109.

Agasimani, C. A., & Hosmani, M. M. (1989). Response of groundnut to FYM, nitrogen and phosphorus in rice fellow in coastal sandy soil. *Journal of Oilseeds Research, 6*(1), 360–363.

Ahmad, N., Mohammad, R., & Khan, U. (2007). Evaluation of different varieties seed rates and row spacings of groundnut planted under agro-ecological conditions of Malakand Division. *J. Agron, 6,* 385–390.

Alam, A. T. M. M. (2002). Yield and quality of groundnut as affected by hill density and number of plants per hill. Pakistan *Journal of Agronomy, 1,* 74–76.

Anandswarup, A., Reddy, D., & Prasad, R. N. (1998). Long-term soil fertility management through integrated plant nutrient supply. Bulletin, All-India Coordinated Research Project on Long-term fertilizer experiment, Indian Institute of Soil Science, Bhopal.

Anonymous "Palm kernel meal as feed for poultry, 1. Composition of palm. Journal of Animal feed science, 2015c. Retrieved 25 May 2015.

Anonymous. "How peanut are grown – Harvesting – PCA". Peanut Company of Australia. 2008.

Anonymous. "Meds & Food for Kids: Medika Mamba". Mfkhaiti. Org. 2015b.

Anonymous. "Nutrition facts for oil, peanut, salad or cooking, USDA Nutrient Data". Conde Nast, USDA National Nutrient Database, version SR-21. 2014.

Anonymous. "Production and trade data for groundnuts (peanuts)". FAOSTAT, Food and Agricultural Organization of the United Nations, Statistics Division. 2015a.

Anonymous. 7CFR 2011: Part 996a. 2011.

Arnon, D. I. (1949). Copper enzymes in isolated chloroplasts, polyphenol oxidase in *Beta vulgaris. Plant Physiol, 24,* 1–15.

Azu, J. N., & Tanner, J. W. (1978). Effect of plant density on growth yield and grade of Spanish peanuts. *Proc. Amer. Peanut Res. Edu. Asso., Inc., 10*(1), 72.

Babiker, E. A. (2004). Effect of sowing date and intra-row spacing on growth and yield on groundnut. *Gezira Journal of Agricultural Science (Sudan) 2,* 26–36.

Ball, S.T., Wynne, J. L., Elkan, G. H., & Schneeweis, T. J. (1983). Effect of inoculation and applied nitrogen on yield, growth and nitrogen fixation of two peanut cultivars. *Field Crops Res. Int., J. 6,* 85–91.

Bandyopadhyay, A., Ghosh, P. K., & Mathur, R. K. (2000). Groundnut situation in India. The present scenario and future strategies. *Indian farming, 50,* 13–20.

Bao, Y., Jiali, J. H., Frank, B., Edward, G. L., Stampfer, M. J, Willett, W. C., et al. (2013). "Association of Nut Consumption with total and cause-specific mortality," *England Journal of Medicine, 369,* 2001–2011.

Belletini, N. M. T., & Endo, R. M. (2001). Compartmento doamendvin (*Arachis hypogaea* L.) sob differentes (e) densidedes (di) sameadura. Acta scientiatium *Universidade-Estadual de-marniga, 23*(5), 1249–1256.

Bhan, S., & Misra, D. K. (1970). Water utilization by groundnut (*Arachis hypogaea* L.) as influenced due to plant population and soil fertility levels under arid zone conditions. *Indian J. Agron, 15,* 258–263.

Bhattacharya, P., & Chakraborty, G. (2005). Current status of orgnic farming in India and other countries. *Journal of Fertilizers, 1*(9), 111–123.

Bhattacharya, P., & Gehlot, D. (2003). Current status of organic farming at international and national level. Agrobios farming at international and national level. *Agrobios Newsletter, 4,* 7–9.

Bhattacharya, P., & Ghosh, P. (2001). Phosphorus use efficiency in brinjal with FYM and sulphur. *Journal of Indian Society of Soil Science, 49* (3), 456–462.

Bheemaih, G., Subrahmanyam, M. V. R., Ismail, S., Sridevi, S., & Radhika, K. (1999). Effect of integrated application of green leaf manures and fertilizers on growth and yield of summer groundnut under different cropping systems. *Indian J. Agric. Sci., 69* (10), 735–737.

Björkman, O., Badger, M. R., & Armond, P. A. (1980). Response and adaptation to temperature. p. 239–249. In: N.C. Turner, and P.J. Kramer (Ed.). Adaptation of plants to water and high temperature stress. John Wiley & Sons, New York.

Boote, K. J. (1982). Growth stages of groundnut. *Peanut Science, 9,* 35–39.

Carley. D. S., Jordan, D. L., Dharmasri, L. C., Sutton, T. B., Brandenburg, R. L., & Burton, M.G. (2008). Peanut response to planting date and potential of canopy reflectance as an indicator of pod maturation. *Agron. J., 100,* 376–380.

Chavan, A. P., Jain, N. K., & Mahadkar, V. V. (2014). Direct and residual effects of fertilizers and biofertilizers on yield, nutrient uptake and economies of groundnut (*Arachis hypogaea* L.) – rice (oryzasiliva) cropping system. *Indian J. Agron, 59*(1), 53–58.

Chawale, V. V., Bharad, G. M., Kohle, S. K., & Nagdeva, M. B. (1995). Effect of N and FYM on yield, quality and nitrogen uptake of summer groundnut under microsprinkler irrigation. *Research Journal, Punjabrao Krishividyapeeth, 19*(2), 171–172.

Das, P. C. (1997). *Oilseeds Crops of India,* Kalyani Publishers, Ludhiana, Punjab, India. pp. 80–83.

Devakumar, M., & Giri, G. (1998). Influence of weed control and doses and time of gypsum application of yield attributes, pod and oil yields of groundnut (*Arachis hypogaea* L.). *Indian Journal of Agronomy 43*(3), 453–458.

Devlin, R. M., & Witham, F. H. (2005). "Plant Physiology", 5th edition, CBS Publishers & Distributors, Delhi (India) PP. 140.

Dosani, A. A. K., Talashikar, S. C., & Mehta, V. B. (1999). Effect of poultry manure applied in combination with fertilizers on the yield, quality and nutrient uptake of groundnut. *Journal of Indian Society of Soil Science, 47*(1), 166–169.

Dutta, D., & Mondal, S.S. (2006). Response of summer groundnut (*Arachis hypogaea* L.) to moisture stress, organic manure and fertilizer with and without gypsum under lateritic soil of West Bengal. *Indian Journal of Agronomy, 51*(2), 145–148.

F.A.O., Food and Agricultural Organization, Special report, 2004.

Fida, K. (2000). Effect of soil amendments on the varietial evaluation of groundnut under rainfed conditional Peshawar valley. Peshawar (Pakistan) NAUP, p. 82.

Fiske, C. S., & Subba Row, Y. (1925). The colorimetric determination of phosphorus. *J. Biol. Chem., 66*, 375–400.

Fortanier, E. J. (1957). De beimloeding van de bloei bij *Arachis hypogaea* L. Doctoral Thesis of Agric., *Univ. of Wageningen, 57*, 1–116.

Gill, O. P., & Kumar, Y. (1995). Effect of sowing dates and varieties on the pod yield of irrigated peanut in Rajasthan, India. In: Proceedings of 82nd Indian Congress, Part IV Late Abstracts, 12, pp. 124.

Giller, P., & Morvan, B. A. (1984). In: "Plant analysis in the control of the nutrition and tropical plants", groundnut (leaf analysis), Martin – Parvel, P., Gagnard, J., & Gautier, P. (Eds.), Lavoisier, Paris (France), pp. 549–558.

Golakiya, B. A. (1988). A system approach to resolve P x Zn antagonisim in different soil water conditions. PhD Thesis, Gujarat Agricultural University, Sardar Krushinagar.

Hadwani, G. J., & Gundalia, J. D. (2005). Effect of N.P. and K levels on yield, nutrient content, uptake and quality of summer groundnut grown on typical haplustepts. *J. Indian Soc. Soil Sci., 53*, 125–128.

Halward, T. S., Tom, S., Elizabeth, L., & Kochert, G. (1992). "Use of single-primer DNA amplifications in genetic studies of peanut (*Arachis hypogaea* L.)" Plant Molecular Biology *18*, 315–325.

Hasan, A. (2009). Effect of diffrent sources of sulphur and Boron on growth, yield and quality of *Brassica juncea* L. (Mustard). PhD Thesis, M.J.P. Rohilkhand University, Bareilly.

Hasan, A., & Abbas, Z. (2007a). Effect of different concentrations of some leaf-applied agro-chemicals on growth, yield and quality in mustard (*Brassica juncea* L.). Paper presented at Indo-Hungarian Workshop, held on 24 March, 2007 at I.V.R.I., Bareilly, sponsored by DST, New Delhi.

Hasan, A., & Abbas, Z. (2007b). Nutritional aspects of some selected crop plants. Inc. "Plant Physiology – current trends" (Ed.) P.C. Trivedi, Pointer Publishers, Jaipur, pp. 153–203.

Hasan, A., & Abbas, Z. (2008). Studies on the effects of Agrochemicals, boron and Sulphur on growth and quality of mustard (*Brassica juncea* L.). In: "Plant Physiology in Agriculture and Forestry" (Ed.) P.C. Trivedi, Aawishkar Publishers, Jaipur pp. 16–56.

Hasan, A., Kumar, A., & Abbas, Z. (2009). Effect of different sources of sulphur and boron on yield characteristics and oil yield in *Brassica juncea* L. *Life Science Bulletin, 6*(2), 233–240.

Hasan, A., Kumar, A., Kumar, P., & Abbas, Z. (2007). Effect of nitrogen levels on growth, herb yield and essential oil content of *Ocimum basilicum* var. *garbratum* (sweet basil). *Indian J. Trop. Biodiv., 15*(2), 140–143.

Hatam, M., & Abbasi, G. Q. (1994). History and Economic importance of groundnut (*Arachis hypogaea* L.) In: "Crop Production" Bashir, E., & R. Bantel (Eds.), Pub. NBF, pp 350–351.

Hirano, S., Shima, T., & Shimada, T. (2001). "Proportion of aflatoxin B1 contaminated kernels and its concentration in imported peanut samples" *Shokuhin Eiseigaku Zasshi., 42,* 237–242.

Holley, K. T., & Hammons, R. O. (1968). Strain and seasonal effects on peanut characteristics, *Univ. Georgia Agri. Expt. Stn. Res. Bull., 32,* 1–27.

Husted, L. (1936). "Cytological studies on the Peanut, Arachis, vol. II", Cytologia, *7,* 396–423.

Ismail, S., Melewar, G. U., Rege, V. S., & Yelvikas, N. V. (1998). Influence of FYM and gypsum on soil properties and yield of groundnut grown in vertisols. *Agropedology, 8,* 73–75.

Jambhekar, H. A. (1992). "Use of earthworm as a potential source to decompose of wastes" Proc. National Seminar on Organic Farming MPKV (PUNE),pp. 52–53.

James, G., & Hasibuan, M. A. (2002). "The composting of sugarcane factory byproducts" Sugarcane May/June (2002) pp. 24–30.

Kachot, N. A., Malavia, D. D., Solanki, R. M., & Sagarka, B. K. (2001). Integrated nutrient management in rainy-season groundnut (*Arachis hypogaea*). *Indian Journal of Agronomy, 46*(1), 516–522.

Kadam, U. A., Pawar, V. S., & Pardeshi, H. P. (2000). Influence of planting, layouts, organic manure and levels of sulphur on growth and yield of summer groundnut. *J. Maharashtra Agric. Univ., 25*(2), 211–213.

Kanaujia, S. N. (2008). Effect of different sources of soil and leaf applied (352) zinc and sulphur nutrients on growth and oil yield of Japanese mint (*Mentha arvensis* L.). PhD thesis, M.J.P. Rohilkhand University, Bareilly.

Kanwar, J. S., Nighawan, H. L., & Raheja, S. K. (1983). Groundnut Nutrition and Fertilizer Response in India. Pub. ICAR, New Delhi.

Karpovikas, A., Gregory, A., & Walton, C. (1994). Taxonomia del genero Arachis (Leguminosae) Bonplandia, *8,* 1–186.

Karpovikas, A., Gregory, A., & Walton, C. (2007). Taxonomy of the genus Arachis (Leguminosae), IBONE, 16 (Supl), 1–205.

Ketring, D. L. (1984). Temperature effects on vegetative and reproductive development of peanut. *Crop Sci., 24,* 877–882.

Kishor, B. (2006). Effect of pyridoxine and nitrogen on growth, yield, essential oil and Biochemical components of *Mentha piperita* L. under salt stress. PhD thesis, M.J.P. Rohilkhand University, Bareilly.

Kishor, B., & Abbas, Z. (2003). Effect of pyridoxine on the growth, yield, oil content and amino: nitrogen content of *Mentha piperita* L Proc. Nat. Symp. on "Plant Biology and Biodivesity in changing Environment", held on Dec 29–31, Jamia Hamdard University, New Delhi, p. 58.

Kishor, B., Kanaujia, S. N., & Abbas, Z. (2006). Effect of vitamin B_6 (Pyridoxine) on the growth, yield, oil content and biochemical components of peppermint *Mentha piperita* L. *Nat. Jour. Life Sci.,* 3(3), 215–220.

Kochert, G., Thomas, S. H., Marcos, G., Leticia, G., Romero, L. C., & Moore, K. (1996). RFLP and Cytogenetic Evidence on the origin and Evolution of Allotetraploid Domesticated peanut, Arachis hypogaea (Leguminosae) Am. *J. Bot., 83,* 1281–1291.

Kumar, A. (2009). Effects of soil-applied rice bran and foliarly-applied its extracts on carbohydrate and protein metabolism, during ripening and maturity in *Saccharum officinarum* L., PhD Thesis, M.J.P. Rohilkhand University, Bareilly.

Kumar, A. (2012). Effect of different sources and methods of nitrogen applications, seed rate and dates of sowing on growth, yield and quality of *Ocimum basilicum* L. (sweet basil) and *Arachis hypogaea* L. (groundnut), PhD Thesis, M.J.P. Rohilkhand University, Bareilly, India.

Kumar, A., Kumar, A., & Abbas, Z. (2014). Effect of pyridoxine and gibberellins sprays on growth, yield and quality of sugarcane (*Saccharum officinarum* L.) *J. Biol. Chem. Research., 31*(1), 349–354.

Kumar, A., Kumar, A., Kumar, P., & Abbas, Z. (2009b). Effect of soil-applied rice bran on germination percentage and cane yield of ten sugarcane (*Saccharum officinarum* L.) cultivars. *Life Science Bulletin, 6*(3), 161–164.

Kumar, A., Sajjad, A., Kumar, A., Verma, N., & Abbas, Z. (2011). Effect of different transplanting dates on herb and oil yield content and uptake of plant nutrients in (*Ocimum basilicum* L.) (Sweet basil). *J. Func. Env. Bot., 1*, 119–121.

Kumar, A., Shukla, S. K., & Abbas, Z. (2009a). Effect of soil-applied rice bran and pyridoxine soaking on sugarcane yield. *Sugarcane International (The Journal of Cane Agriculture), U.K., 27*(5), 210–211.

Kumar, A., Verma, N., Kumar, A., & Abbas, Z. (2010). Crop growth period and population density variations in relation to leaf nitrogen, floral initiation and pod yield of groundnut (*Arachis hypogaea* L.), Life Science Bulletin, 7(2), 193–195.

Kumar, S. (2008). Salinity induce percent seed germination and seedling growth, oil yield of sunflower (*Helianthus annus* L.). PhD thesis, M.J.P. Rohilkhan University, Bareilly.

Kumar, S., & Abbas, Z. (1992). Studies on different sources and methods of zinc application on wheat and triticale. *Geobios, 19*, 156–159.

Kumar, S., Ahmad, A., & Masood, A. (2007). Salinity induced percent seed germination and seedling growth of Brassica napus cv. Agami. Indian *J. Trop. Biodiv., 15*, 90–93.

Kumar, S., Ahmad, A., & Masood, A. (2008a). Salinity induced percent seed germination and seedling growth, oil yield of sunflower (*Helianthus annuus* L.). cv. PAC-3776. *Research on Crops, 9*, 274–277.

Kumar,Y., Shaktawat, M. S., Singh, S., & Gill, O. P. (2003). Effect of sowing date and weed control methods on yield attributes and yield of groundnut (*Arachis hypogaea* L.) *Indian J. Agron., 48*, 56–58.

Lagatu & Maume (1930, 1934) In: Plant Physiology – A Treatise, vol. III. Steward, F. C. (ed.), Academic press, New York, 1963.

Latha, H. S., & Sharanappa (2014). Production potential, nutrient use efficiency and economics of groundnut (*Arachis hypogaea* L.) and (*Allium cepa*) cropping system under organic nutrient management. *Indian J. Agron., 59*(1), 59–64.

Lindner, R. C. (1944). Rapid analytical methods for some of the more common inorganic constituents of plant tissue. *Plant Physiol., 19,* 76–89.

Lourduraj, C. A. (2000). Effect of irrigation and manure application on the growth and yield of groundnut. *Acta Agronomica Hungarica, 18*(1), 83 –88.

Lourduraj, C. A., Geethalakshmi, V., Devsnapathy, P., & Nagarajan, P. (1996). Drought management in rainfed groundnut. *Madras Agriculture Journal, 83*(4), 265–266.

Lundegardh, H. (1951). "Leaf analysis," Translated by R.L. Mitchel, Hilger Division, London.

Maity, S. K., Giri, G., & Deshmukh, P. S. (2003). Effect of phosphorus, sulphur and planting methods on growth parameters and total yield of groundnut (*Arachis hypogaea* L.) and sunflower (*Helianthus aunus* L.) Indian J. *Plant Physiol, 8*(4), 377–382.

Manoharan, V., Sethapathi, R., Ramalingam., & Sivaran, M. R. (1988). Correlation studies in Virginia bunch groundnut (*Arachis hypogaea* L.) Indian Journal of Oilseed Research, *5*(2), 150–152.

Mishra, C. M. (2001). Effect of farmyard manure and chemical fertilizers on the yield and economics of groundnut (*Arachis hypogaea* L.) under rainfed condition. *Madras Agriculture Journal, 87*, 517–518.

Mohapatra, A. K. B., & Dixit, L. (2010). Integrated nutrient management in rainy season groundnut (*Arachis hypogaea* L.). *Indian J. Agron, 55*(2), 123–127.

Moretzsohn, M. C., Gouvea, E. G., Inglish, P. W., Soraya, L., Valls, J. F. M., & David, J. (2013). "A study of the relationship of cultivated peanut (*Arachis hypogaea* L.) and its most closely related wild species using intron sequences and microsatellite markers". *Annals of Botany, 111*, 113–126.

Murthy, S. K., Krishna, & Rao, Yogeswara (2000). Correlation between weather parameters at different phenophases and growth and yield parameters of groundnut (*Arachis hypogaea* L.), *Annals of Arid Zone, 39*, 29–33.

Nambiar, K. K. M., & Ghosh, A. B. (1984). Long term fertilizer experiment, Research Bulletin, Indian Agricultural Research Institute, New Delhi, pp. 101.

Nautiyal, P. C., Ravindar, V., & Joshi, Y. C. (1999). Net photosynthetic rate in peanut (*Arachis hypogaea* L.) Influence of leaf position, time of day and reproductive sink. *Photosynthetica, 36*, 129–138.

Nawlawade, P. P., & Patil, B. P. (2000). Sowing time and seed bed modification for yield maximization in groundnut in north Konkan. *Journal of Agricultural Meteorology, 2*, 152–157.

Ozcan, M. M. (2010). "Some nutritional characteristics of kernel and oil of peanut (*Arachis hypogaea* L.)". *J Oleo Sci., 59*, 1–5.

Pacharne, D. P., Tumbare, A. D., & Dhonde, M. B. (2016). "Productivity, profitability and nutrient uptake in groundnut (*Arachis hypogaea*)-based cropping systems under different nutrient-management practices. *Indian J. Agron., 61*, 161–167.

Palaniappan, S. P., & Annadurai, K. (1985). In: Organic -farming-theory and practice, pp. 53–73.

Panchaksharaiah, S. (1999). Effect of nitrogen and iron on growth and yield of groundnut. *J. Farming Systems, 1*, 9–13.

Panse, V. G., & Sukhatme, P. V. (1967). *Statistical Methods for Agricultural Workers,* ICAR, New Delhi. pp. 137–174.

Panwar, A. S., & Singh, N. P. (2003). Effect of conjunctive use of phosphorus and bioorganics on growth and yield of groundnut (*Arachis hypogaea* L.). *Indian J. Agron., 48*(3), 214–216.

Patel, M. S., Gundalia, J. D., Davaria, R. L., & Patel, B. D. (1994). Effect of potassium and hormones under normal and higher levels of N and P on yield and uptake of nutrients by summer groundnut Gujrat Agricultural *University Research Journal, 20*, 23–38.

Patra, A. K., Tripathi, S. K., & Samui, R. C. (1995). Physiological basis of yield variation in rainfed groundnut. *Indian J. Plant Physiol., 38*, 131–134.

Ramesh, R., Shanthamulliah, N. R., Jayadeva. H. H., & Hiramath, R. R. (1997). Growth and yield of irrigated groundnut as influenced by phosphorus farmyard manure and phosphate solubilizing micro organism. *Current Research, 26*(1), 196–198.

Ramesh Babu, N., Reddy, S. R., & Raghaiah, R. V. (1985). Studies on nutrient utilization by groundnut. *Indian J. Agron., 30*, 120–121.

Rao, S. S., & Shaktawat, M. S. (2001). Effect of organic manure, phosphorus and gypsum on growth, yield and quality of groundnut (*Arachis hypogaea* L.). *Indian J. Plant Physiol, 6*(3), 306–311.

Rao, S. S., & Shaktawat, M. S. (2002). Effect of organic manure, phosphorus and gypsum on groundnut (*Arachis hypogaea* L.) production under rainfed condition. *Indian J. Agron., 47*(2), 237–241.

Ravi Kumar, A., Raghavalu, P., Reddy, G. V., Subbaiah,G. V., & Rao, G. V. H. (1994). Effect of dose and time of nitrogen and gypsum application on dry matter production and yield of groundnut (*Arachis hypogaea* L.). *Indian J. Agron, 39*, 323–325.

Ray, S. P., Chaudhary, B. P., & Chaudhary, I. B. (1997). A report on outreach research on groundnut (1996/97), Sarlahi (Nepal). Oilseed Research Programme (n.p.).

Reddy, B. S., Reddy, S. R., & Subbiah, G. (1981). Effect of supplemental nutrition during post flowering phase of groundnut. *The Andhra Agric. J., 38*, 4–7.

Reddy, M. S. S., Reddy, S. R., Reddy, M. G., & Reddy, M. G. R. K. (1991). Residual effect of organic matter and fertilizer applied to groundnut and maize on succeeding groundnut and groundnut-maize- groundnut sequence. *Indian J. Agron, 86*, 298–300.

Reddy, S. R. (2004). "Groundnut": In: *Agronomy of Field Crops*, Chapter 9. Kalyani Publishers, New Delhi, pp. 381–419.

Sajjad, A., Kumar, A., Kishor, B., & Abbas, Z. (2011). Morphophysiological performance of seedling of three varieties of chickpea (*Cicer arietinum* L.). *Life Science Bulletin, 8*(1), 79–81.

Salama, N. E., Hanna, F. R., & Ahmad, A. (1994). Flower production and yield of groundnut under various concentrations of organic manure and water amounts. *Ann. Agric Sci. (Egypt) 32*(1), 1–19.

Salisbury, F. B., & Ross, C. N. (2000). *Plant Physiology*, 5th Indian Reprint, Cengage Learning India Pvt. Ltd., New Delhi.

Sarawagi, S. K., Tiwari, S. K., & Tripathi, R. S. (1998). Nitrogen fixation, balance sheet and yield of winter soyabean as affected by divergent nutrients. *Annals of Agricultural Research, 9*(4), 379–385.

Sarkar, R. K., Chakraborty, A., & Bala, B. (1998). Analysis of growth and productivity of groundnut (*Arachis hypogaea* L.) in relation to micro nutrient application. *Indian J. Plant Physiol, 3*(3) (N.S.) 234–236.

Seijo, G., Graciela, I. L., Fernandez, A., karpovickas, A., Ducasse, D. A., David, J., et al. (2007). 'Genomic relationships between the cultivated peanut (Arachis hypogaea, Leguminosae) and its closed relatives revealed by double GISH'. *Am. J. Bot., 94*, 1963–1971.

Selamat, A., & Gardner, F. P. (1985). Growth, Nitrogen uptake, and partitioning in nitrogen fertilized nodulating and non nodulating peanut. *Agron. J., 77*, 862–867.

Sharma, A. (2007). Ethno-botanical studies on the Tharu tribe of Udham Singh Nagar, Uttaranchal. Ph. D., thesis, M.J.P. Rohilkhand University, Bareilly.

Shukla, S. K. (2010). Effect of some agro-chemicals on sugar accumulation and quality in early and late maturing sugarcane cultivars. PhD Thesis, M.J.P. Rohilkhand University, Bareilly.

Sims, J. T. (1986). Phosphorus fractionation, adsorption and description studies with-heavily manured and fertilized atlantic coastal plain soils. Agronomy Abstracts p. 172 Fidc. Advances in Agronomy 52, 1994.

Singh, B., & Singh, N. Y. (2002). Concepts in nutrient management. In: Recent Advances in Agronomy (Singh, G., Kolar, J. S., & Sekhon, H. S. Eds.) pp. 92–109. Indian Society of Agronomy, New Delhi.

Singh, K. P., & Ahuja, K. N. (1984). Dry matter accumulation, oil content and nutrient uptake in groundnut (*Arachis hypogaea* L.) cv. T64 as affected by fertilizers and plant density. *Indian J. Agron., 30*, 411–45.

Singh, S., Kumar, Y., & Gill, O. P. (2003). Growth characteristics of summer groundnut (*Arachis hypogaea* L.) as influenced by sulphur levels, irrigation schedules and organic manures. *Annals of Biology (Hissar), 19*(2), 135–139.

Singh, S., Singh, A. L., Kalpana, S., & Misra, S., Genetic Diversity for growth, yield and quality traits, *15*(3), 267–271.

Singh, S. B., Singh, M. P., Matamber., & Nidhi, Katiyar. (2006). Microbial composting of pressmud and its impact on soil productivity. "Seminar on efficient practices in sugarcane" held on 25 Dec U.P. Council of Sugarcane Research, Shahjahanpur. U.P. India 2006.

Singh, S. R., Nayar, G. R., & Singh, U. (2007). Productivity and nutrient uptake of soyabean (*Glycine max*) as influenced by bio-inoculants and farmyard manure under rainfed conditions. *Indian J. Agron., 52*(4), 325–329.

Steward, F. C. (1963). Plant Physiology (Ed) vol. III, A treatise, Academic Press, New York.

Steward, F. C., & Durzan, D. J. (1965). Metabolism of nitrogenous compounds. In: Steward, F.C. (Ed.). *Plant Physiology: A treatise*. Vol. IV, p. 381, Academic Press, New York, USA.

Suman, A., Lal, M., Singh, A. K., & Gaur, A. (2006). Microbial biomass turnover in Indian sub-tropical soil under different sugarcane intercropping systems. *Agron. J., 98*, 698–704.

Survace, D. N., Dongale, J. H., & Kadrekar, S. B. (1986). Growth, yield, quality and composition of groundnut as influenced by FYM, calcium, sulphur and boron in lateritic soil. *J. Maharashtra Agric. Uni., 11*, 49–51.

Talwar, H. S., Nagaswar Rao, R. C., & Nigam, S. N. (2002). Influence of canopy attributes on the productivity of groundnut. *Indian J. Plant Physiol., 7*(3), 215–220.

Taufiq, A., & Sudaryono. (1998). Sulphur and organic manure fertilization on groundnut in high pH of alfisol Balai penelition Tanaman Kacang Kacangan dan umbi umbian, Malang (Indonesia), Bali kabi, pp. 198–208.

Taylor, C. L. (2003). "Qualified Health Claims": Letter of Enforcement Discretion – Nuts and Coronary heart disease (Docket No 02P-0505). Center for Food Safety and Applied Nutrition, FDA.

Trivedi, B. S., Bhatt, P. M., Patel, J. M., & Gami, R. C. (1995). Increased efficiency of fertilizers through addition of organic amendment in groundnut. *Journal of Indian Society of Soil Science, 43*(4), 627–629.

Varalakshmi, L. R., Srinivasa, Murthy, C. A., & Bhaskar, S. (2005). Effect of integrated use of organic manures and inorganic fertilizers or organic carbon available N.P.K.

in sustaining productivity of groundnut finger millet cropping system. *J. Ind. Soc. Soil Sci., 53*, 315–318.

Venkaiah, K. (1985). Effect of bulk density of growth and yield of groundnut. *Indian J. Agron., 30*, 278–280.

Verma, N. (2011). Studies on balanced nutrition integrated with organic manures in groundnut (*Arachis hypogaea* L.) PhD Thesis, M.J.P. Rohilkhand University, Bareilly.

Verma, P. K., Sharma, G. D., Dhindsa, K. S., & Sangwan, N. K. (1984). Effect of spacing on growth attributes and oil contents in palmarosa (*Cymbopogon martinii* var. motia) grass Indian J. *Plant Sci., 2*, 45–49.

Verma, V. (2007). "Textbook of Plant Physiology" New Edition, Emkay pub. Books India, New Delhi, pp. 211–212.

Went, F. W. (1953). The effect of temperature on plant growth. *Ann. Rev. Plant Physiol., 4*, 347–362.

Wood, I. M. W. (1968). The effects of temperature at early flowering on the growth and development of peanuts. *Aust. J. Agric. Res. 19*, 241–251.

Wynne, J. C., Ball, S. T., Elkan, G. H., & Schneeveis, T. J. (1979). Cultivar, inoculum, and nitrogen effects on nitrogen fixation of peanuts. Agron. Abstr. American Society of Agronomy, Madison, W.L.

Yadav, D. V. (2001). "Recycling of sugar factory pressmud in agriculture". Proc. Seminar on sugarmill waste based vermicompost and biofertilizers Organized by U.P. Ganna Kisan Santhan 11 Butler Road, Tilak Marg, Lko., held on 30 Sept–01 October pp. 91-106 2001.

Yakabri, M., & Satyanarayan, V. (1995). Dry matter production and uptake of nitrogen, phosphorus and potassium in rainfed groundnut. *Indian J. Agron., 40*, 325–327.

INDEX

Milton Keynes UK
Ingram Content Group UK Ltd.
UKHW022047141024
449569UK00022B/839

9 781774 636343